THE CAT WHISPERER

THE CAT WHISPERER

THE SECRET OF HOW
TO TALK YO YOUR CAT

CLAIRE
BESSANT

JOHN BLAKE

Published by John Blake Publishing Ltd
3 Bramber Court
2 Bramber Road
London W14 9PB, England

First published in hardback in 2001

ISBN 1 903402 42 5

British Library Cataloguing-in-Publication Data: A catalogue
record for this book is available from the British Library.

Typeset by GDAdesign
Set in Goudy

Printed in Great Britain by Creative Print and Design
(Wales), Ebbw Vale, Gwent

9 10 8

Papers used by John Blake Publishing Ltd are natural,
recyclable products made from wood grown in sustainable forests.
The manufacturing processes conform to the environmental
regulations of the country of origin.

For Steve

Contents

Prologue

We have come a long way in our understanding of animals in recent years. Documentaries on wild animals have given us an insight into their natural behaviours – the reasons why they do what they do. And funnily enough, looking at the behaviour of wild animals has brought us around to looking at the behaviour of our own pet animals. This was something which was taken very much for granted for many years – wild animals were interesting; domestic animals just did our bidding. Now we know that looking at both wild and 'domesticated' animals can help us to understand both better.

Pets have shared our lives for thousands of years, yet not even 20 years ago (and sometimes still now) we were using very punitive methods of dog training. Remember the first dog training programmes on the television? As a nation we were riveted and all went out and bought choke chains and said 'sit' to our dogs in high-pitched voices with the appropriate hand signals. It was still a very dominating type of training and punishment figured quite highly.

Things then moved on when a new generation of 'animal behaviourists', rather than 'trainers', began to take an interest in behaviour problems in pets and in helping owners with dogs which were nervous or aggressive, noisy or pulled on the lead. We began to look at the dog as genetically similar to a wolf and to compare the behaviour we saw on documentaries to those in our pet dogs. This allowed behaviourists to start to explain the motivations and thus perhaps the solutions to 'behaviour problems'.

The recent Horse Whisperer phenomenon has shown us that years of 'breaking' horses by forcing our will on them is not the best way to go about getting this very large and powerful creature to work with us. It took one man called Monty Roberts to study how horses behaved in the wild and how they interacted with one another to show us that there was a different and very successful way to work with horses. Horses can actually choose to be with us. He used the body language of the horses and a knowledge of how they bond with one another and applied it to the horse/man situation with astounding results. Anyone who has seen his method of 'joining up', where a horse is made to gallop around a ring without any type of restraint, initially shying away from any type of contact and yet within a few minutes of the man using the right body language and picking up the horse's body language, the horse begins to follow him around with his head on his shoulder, knows what a highly emotional experience it is. There is no forcing or cajoling – just a willingness on the part of the horse to be with the man. For those who respect and love animals there can be no bigger thrill than one coming to you of its own free will, understanding what you are trying to communicate, and trusting you not to harm it.

Horses and dogs both originate from group-living animals – their behaviours have developed because they have to communicate with that group and staying within it means protection and all the other benefits of companionship. To some extent this ability to fit into a group has meant that man has been able to force them to join into the human group and so, before we learned about communication and using natural behaviour and rewards, coersion worked to some extent. However, the cat is a different kettle of fish altogether, as any 'feline-o-phile' will tell you.

The cat has altered very little from its wild ancestor – the African Wildcat. It is a solitary hunter and doesn't need to have any other cats around. It may enjoy some social interaction but that is by choice rather than necessity. Anyone who has tried to get a cat to be obedient using the traditional dog-training methods will have had no success whatsoever: there is no motivation for the cat to stay around if it is not enjoying the relationship – it has no need of a pack or herd to give it support. Yet cats have lived around us since Egyptian times and it has been a mutually beneficial relationship – the cats dealt with vermin and we enjoyed the cats' company. However, we have never been able to mould them as we have dogs – neither physically nor mentally. The cat has remained very independent, yet has moved more and more into our homes and our lives. We have mostly taken this for granted.

Then we started to look at the cat's behaviour in the wild; not only its wilder cousins, but feral cats or even pet cats outside in the garden rather than inside or on our laps. Like the Horse Whisperer, people began to piece together what a cat really was – not a purring fluffy baby, but an astonishing hunter, an animal with very interesting communication techniques and an ability to transfer from being a cat's cat to a person's cat without losing any of its dignity or abilities.

Inevitably we have forced the cat to live closer and closer to us in modern lives, and to share territory with a great many other cats in our urban and suburban homes. It has coped well, with the occasional problems such as marking the indoor of the house as well as the outdoor territory, or scratching the furniture instead of the trees. The latest generation of feline behaviourists know that to understand your cat and enjoy him to the full it helps greatly to know what motivates him and to be able to give him space and the chance to carry out his

3

natural behaviours within the limits of our 'ownership' as well as security within his indoor territory. Cat whispering certainly helps when problems arise too – most problems are natural behaviour occurring in the 'wrong' place (well, wrong to us humans) but they are a signal from the cat that things are not all right. Like the Horse Whisperer we need to look at the natural cat and reach out to it so that it is not frightened of us and can find a way to change its behaviour within our homes so that harmony is restored once again.

How we see our animals and how we can help them when they are feeling unsure of their place within our homes has come a long way; it is still changing as we learn more, but the scene has been set – the animal whisperer is here to stay and we should all look forward to more insights into what makes our pets tick and to a more fulfilling relationship with them.

This book sets out to explain how cats function as cats and how that can be successfully transferred and used in building the cat/human bond both on a day-to-day basis and when problems occur. Find out more about your cat and you will appreciate it even more, not only for its behaviour and abilities, but for the fact that it can share its life with us in such a smooth and rewarding manner. Get whispering.

1

A Different World

Imagine trying to communicate with an Eskimo from the frozen world of the Arctic if you have only ever lived in the African jungle. Not only are there language differences, but many of the common references of everyday life such as landscape, vegetation, wildlife, dress – or lack of it – and daily hazards and routines are extremely varied. Although the aims of having a shelter and keeping the body fed and at a comfortable temperature are similar, how each is achieved and their relative importance to survival are not.

When trying to understand how and why members of a completely different species act as they do, an insight into how they see their world and an examination of some of the problems they have to overcome can be very enlightening.

We tend to assume everything revolves around how we tall, rather stiffly upright and slow humans move through our world with good eyesight (so long as there is daylight), hearing adequate to receive the sounds of other humans and a relatively poor sense of smell. Although it lives in the same physical environment as we do, the cat may see things in a very different way and, by using its heightened senses of hearing, smell and touch, be influenced by factors of which we are only vaguely aware. Built as a perfect hunter, its senses attuned for tasks which we are swift or silent enough to undertake only in our imagination, the cat may move in our world, but it is a world we would barely recognise – like an Eskimo who wakes up in the jungle!

Imagine yourself moving at shin level but having the ability to leap (from a standing start) to five times your height with seemingly little effort. Climbing a tree, leaping from a wardrobe or balancing and walking along the top of a very slim fence are not difficult tasks, even for a geriatric cat. With a lithe body and nimble feet which can convert instantly from silent padders to crampons or grasping tools, the cat is equal to most terrains and can survive in conditions from Arctic to desert.

The cat's sense of balance and its ability to fall on its feet have an almost supernatural reputation. Having evolved an automatic sequence of moves which come into play as it falls through the air, a safe landing is almost guaranteed. The cat levels its head, flips its top half round to face the ground and then flips its rear end round too. Using its tail, the cat counteracts any overbalance and can land on all four feet with its back arched to cushion the impact. Of course, this does not mean that cats are never killed or injured in falls – they often are. If they miscalculate and land badly, or fall from a very great height, the impact is likely at least to break their legs, so they are not that supernatural!

This special sense of balance, which is based in the inner ear, may be the reason that cats do not suffer from motion sickness, a feeling of nausea which afflicts both humans and dogs when the head has frequently to change its position relative to the body, such as when riding in a car or sailing in a boat. While cats may not enjoy travelling in the car, they are not usually sick, unlike puppies or young children who seem to be affected more than their adult counterparts.

Supple and rarely clumsy, the cat has a physique that enables it to perform feats way beyond those of even our best athletes. Add to that its lack of predators (apart from some of the human species), its amazing hunting talents and ability to adapt to life wherever it may be, with or without help from us, and you may be on the way to fathoming that inexplicable independence and confident air of the cat.

So how does the cat feel, see and hear and how is its perception of the world so different from that of man?

FEELING THE WORLD

If we draw a picture of a human with the parts of the body large or small, relative to how touch-sensitive those parts are, the distorted image (called a homunculus) has large hands, lips, tongue, genitalia and a relatively small back, legs, feet and arms. If you think about where you 'feel' and how you use your fingertips to, say, try to locate a tiny splinter which you know is in your thumb but you cannot see, you can immediately focus on your sense of touch. If you fail to find the splinter with your fingers, you automatically use that even more sensitive organ, the tip of your tongue. When you draw the same touch-sensitive representation of a cat (what could be called a felunculus), the creature also has a large head (especially the tongue and nose) and huge paws. Watch a cat investigating a new object, learning to play with prey or simply creeping up on something strange. It will first timidly touch the object once with a paw, then repeat the action a little more confidently, and then move in closer to investigate with its nose. The pads are very sensitive to touch and vibration – which is perhaps the reason many cats do not like their pads being stroked. Maybe it feels like 'tickled feet', one of the sensations some humans cannot bear either.

Funnily enough, although sensitive to touch, the pads are not very sensitive to hot or cold things. The nose and upper lip are the only parts of the cat's body which are very sensitive to temperature and it uses these to estimate the temperature of food and the environment. As a tiny kitten its sensitive nose homed in on the warmth of its mother like a little heat-seeking missile. Even at one day old a kitten can detect and move along a thermal gradient, avoiding cold hard surfaces, to reach the warmest, softest spot next to its mother.

While the nose and lips can be used to sense the temperature of the surroundings or of the cat's food, the rest of its body is relatively insensitive to high or low temperatures. This probably explains why so many owners are shocked when their cats leap on to a hot cooker hob with little care, or curl up in the embers of a fire which is not quite dead and singe their fur. Humans will move away from a temperature of 44°C because it feels uncomfortably hot. The cat can happily stay put at way beyond this temperature, up to about 52°C, which explains its ability to sit in front of a roaring fire or on a boiling-hot radiator and still enjoy it.

The cat is also covered with what have been called 'touch spots' (about 25 of them per square centimetre of skin), which are areas of skin rich in touch-sensitive nerves. Spraying a cat with a fine shower of water may make it quiver – its skin is reacting in sequence as the tiny droplets hit each sensitive spot in turn, and the skin literally 'ripples'.

THROUGH CAT'S EYES

Although the structure of the cat's eye is very similar to that of other mammals, it does have its own peculiarities and specialities. It is now thought that the cat does not see only in black and white, but that it can actually detect some colour. By turning down the colour control on the television so that the blues and greens just show, and imagining the dull reds as grey, you can go some way to visualising how the cat sees the world.

Because they hunt at twilight, when most of their rodent prey is on the move, cats do not really need to see colour. Twilight is a very strange type of light in which all colours soften and fade. While it may still seem quite 'light' to us, we are actually very bad at discerning what we are looking at. This explains the increased risk of road accidents at twilight and why we ask people to 'wear something white' in order to be noticed. However, this time of day is when the cat comes into its own.

Although cats may not be able to see the same fine detail as we can in daylight and cannot focus so well on nearby objects (they see best at about two to six metres), when it comes to following movements the cat does not miss a twitch. Special nerve cells in the cat's brain respond to the smallest movement – an obvious advantage to a hunter. The lightning-fast feline reflexes, combined with an ability to judge distance very accurately, allow the cat to home in on its prey with devastating speed. However, if the prey freezes – an action many 'prey' species learn as a defence strategy – the cat may lose sight of it. If the cat has already pinpointed the little creature though, it may perform that familiar little 'wiggle' of its bottom, which may slightly alter the stalker's view and stimulate its movement-sensitive eyes to reassess the prey's position before making a strike.

The cat has very large eyes in relation to the size of its head and part of its appeal is thought to be that it resembles a human baby in this respect. Kittens also, like babies, are born with blue eyes. However, the most amazing feature of the cat's eye must be its adult colour, splendid in whichever hue of yellow, green, blue, lavender or orange. This gleaming colour in the part of the eye called the iris has no function, but like the beautiful paint on a racing car, it merely covers a very special mechanism. The muscles in the iris alter the position of the pupil to allow the correct amount of light to enter the eye. It looks like a curtain across the pupil – when drawn fully across, little light can get through. In full sun you may only see a minute vertical dark strip of the pupil down the centre of the eye; this ensures that the sensitive layer of cells called the retina, at the back of the eye, is protected from over-exposure. When the light is low the iris changes position, allowing the pupil to dilate and more light to enter.

Sundown for the cat does not mean that everything disappears into darkness. Not only can the pupil open up to about one centimetre in diameter (making the whole eye look black), but a special layer of cells behind the retina reflects back any light which has not been absorbed on its way through, so that the eye gets a second chance to intercept the image. This pigment layer of cells, called the tapetum, acts like a mirror and is responsible for the gold or green shine to the eyes which can be seen when a cat is caught in car headlights at night. One of the reasons the Egyptians regarded the cat as sacred was this glowing-eye phenomenon. The strange image we find on photographs taken by flash-light – eyes aglow in fluorescent green instead of their normal warm yellow or orange – is caused by the same reflecting layer. Interestingly the blue eyes of the Siamese cat glow a sort of blood-red colour in flashlight photos because they have a slightly different make-up from other cats in their reflecting layer of cells. Pictures are often taken to capture that very beauty of the eye's true colour, so it can be a frustrating business if one ends up with two glowing green embers. Photographing using natural light outside, in the garden for example, should show the colour of the eye in all its splendour, so long as the cat can be persuaded to remain still.

So not only can the pupil open very wide and get two chances to absorb the light, the eye of the cat also has extra sensitive nerve cells in the retina which respond to the light hitting them. This gives the cat the ability to see in light about six times dimmer than we ourselves need. Cats are able to see in light so weak that many scientific instruments can hardly detect it, and probably twilight to them still seems like daylight does to us.

The eye is not only an amazingly sensitive instrument for the cat, it is also an indicator of mood. Eyes are said to be the windows to the soul, and pleasure, fear or excitement can make the pupil dilate . . . but more of this later.

'SEEING' BY TOUCH

There is more to 'seeing' than simply looking through the eye – as any blind person could tell you. Other senses can also help animals to 'see'. Whiskers are extremely sensitive instruments and function in a type of see/feel way. Look at your cat sideways with the light behind it and not only will the dozen or so whiskers in rows on the upper lip be obvious, but you may also notice some above the eye and on the chin. Similar coarse hairs (actually called vibrissae) are also found on the elbows. These form a sort of 'force field' of sensitivity – they are turned on by movement, even a slight breeze, and can be put 'on alert' when required.

The whiskers sprout from a deeper layer of the skin than other hairs do and they act like levers, magnifying any slight movement while they bend. This stimulates nerve endings which can detect the speed and direction of movement and provide detailed information on the cat's surroundings – they are even thought to enable the cat to sense minute air disturbances which occur around obstacles so it can 'feel' their presence, even when it can't see them. These nerve impulses travel along the same path to the brain as does information from the eyes. The brain then uses the two systems to build up a three-dimensional picture of the environment.

A cat without whiskers may be very unsure when it's moving in the dark or through narrow spaces. In dim light, when the pupil is fully dilated to let in as much light as possible, the eye is less able to focus on close objects – so the cat uses its whiskers to detect the world immediately around it. A touch on the whisker will cause a reflex closing of the eye, ensuring that any twig or grass springing back will not cause injury; this is a vital protective measure for the hunter with its eyes firmly fixed on prey moving through hedges, grass, or around small holes. A cat with poor eyesight moves its head from side to side as it walks, using its whiskers as minesweepers or the equivalent of a blind person's white stick to help it negotiate gaps and obstacles. In the USA some blind dogs have been fitted with 'whiskers' – flexible plastic sticks fixed to the dog's collar on either side of its face. These dogs can now feel their way around their homes with great success, and they have learned both how to avoid obstacles in their path which touch the whiskers and to stop when the whiskers lose contact with the ground, so that they won't fall down a step. The fitting-out with 'whiskers' has given the dogs a new lease of independent life, for which they can thank their feline cousins.

The cat also uses its whiskers to feel its prey as it makes its final hunting pounce and grabs the animal with its teeth and claws. A hunt is a very exciting time for the cat and its body is kept 'primed' by the hormone adrenalin. Adrenalin also causes the pupils in the eye to dilate and makes it very difficult for the cat to focus on close-by objects, such as the shrew it has in its mouth. To feel its way to the correct position for the killing bite to the animal's neck, it uses its whiskers like that 'third hand' we would all find so useful.

Have you ever studied your cat's face and noticed, scattered among the rows of whiskers above the upper lip, lots of dark dots which look like rows of beauty spots? These have been studied in lions and researchers found that the positioning and patterns they form are different on each lion, as our fingerprints are unique to each of us. No one has yet studied the patterns in domestic cats (they would have a bit of difficulty with black-haired ones) and anyway we can tell our cats apart by much more obvious clues, but it is an interesting finding.

HEARING

The dog has always been thought of as the keen-eared pet, which belief resulted in a craze for ultrasonic whistles that were blown silently, trusting that at least the dog could hear. The fact that most people reverted to the well-tried methods of shouting or whistling through their teeth probably illustrates our lack of faith in things we cannot actually see or hear, rather than making us admit that our dogs are just badly trained. In fact, when it comes to sound-sensitivity the cat can hear sounds of even higher frequency than the dog can – up to about 60 kHz. Man is sensitive up to only about 20 kHz, which means that we miss many of the high sounds made by small rodents, the primary prey of the cat.

The cat's mobile ears, which can rotate independently through 180°, allow it to collect sound from all around without even moving its head. A cat 'asleep' in the sun will still move its ears like radar saucers, scanning the surroundings for hints of danger or sounds of prey. Thirty muscles (man has only six) control each outer ear – and these funnel sounds down into the inner ear. Locating prey very accurately by sound means that the cat can move in swiftly and directly and not rely solely on its sight to pinpoint the animal's position. This is beautifully illustrated by a small event which took place one summer evening in our garden. Bullet, our black-and-white-ace-hunter moggie, was spotted crossing the lawn with a newly caught mouse. My husband and I rushed out to attempt a rescue and, as we galloped towards him, Bullet dropped the mouse and ran off. We diligently searched the closely cropped lawn for the little creature so that we could check it over and release it in a safe place (if there is such a thing with Bullet on the prowl!). Having the advantage of height and expecting to spot it immediately, we both scanned frantically, bent over ready to scoop the mouse up. Suddenly Bullet rushed in from the sidelines, grabbed the mouse in one graceful movement and tore off into the hedge. The mouse had been right under our noses, yet neither of us could see it. Bullet had spotted and heard its movement from quite a long distance away and captured his prey in one swift sure move, leaving us mad at him for being so clever and mad at ourselves for not seeing it first, or at least dividing our attentions so that one of us shut Bullet in the house,

It is often difficult to know whether to get involved and try and rescue animals or to let nature take its course. Having tried many different approaches with Bullet, who is such a successful hunter, I have come to the conclusion that it is only worth trying to 'rescue' the creature if it has just been caught and is probably not yet badly injured. Bullet often grabs the bird or mouse and comes indoors. The best way I have found to make him drop the animal is to put a cloth over him (usually a tea towel comes to hand first). This is much more successful than getting involved in a physical struggle. The bird then flaps off around the kitchen and has to be rescued from behind the fridge; however, it is usually only in shock and not badly injured and can be released. I'm afraid that my success record for trying to keep and nurture injured birds or small mammals is virtually zero, so if the animal looks as if it is badly injured I do not take it away from the cat but hope that he deals with it quickly, rather than trying to 'nurse' it and in actual fact drawing out its agonies even longer.

Using its senses of touch and hearing, a blind cat will follow a toy along the ground in play and can even hunt successfully. Onlookers would not guess the cat's handicap until the toy is lifted off the ground, when the cat immediately loses its ability to chase because, in the air, the toy causes no vibrations, friction or tiny sounds with which the cat can accurately determine its position. Using this acute hearing, and feeling with its whiskers when in close range, the blind cat overcomes its visual disability quite well. Although the world of the cat is also filled with noises too faint or too high in pitch for us to hear, we can use the appeal of sound to play with our pet felines. They often like games with materials such as cellophane that make high-pitched crackling and squeaking noises similar to those they home in on while hunting.

Because cats have such acute hearing they are able to use a wide vocabulary for communication. They respond to high sounds such as those of kittens squeaking and use many sounds with us and other cats. In 1932 French researchers reported that a certain sharp 'mi' sound is a powerful sexual stimulus for cats – the caterwauling of tom-cats probably excites and deters other toms and makes the queen more receptive to mating. Some cats react when musical instruments are played or respond to certain notes. One scientist trained cats to respond to a whistle pitched at middle C. He found they could distinguish between middle C and whistles as close in pitch as half a tone – an accomplishment many humans could not boast of. Cats come to know familiar sounds, such as their owner's footsteps or car engine and will leap up at the sound of the cutlery drawer opening or the squeak of the tin opener at mealtime. If it reacted to every sound reaching the ear, the cat's brain would be continually overloaded by the huge volume of information and would be unable to select those noises which mean danger, food or challenge. So it has the ability to fade insignificant, familiar noises into the background while still remaining receptive to any strange noise, just as people living near a railway line will not notice a train go by though visitors will wonder how the residents could possibly put up with the constant rumblings. Some white cats are born deaf (an inherited trait linked to their coat colour) and old cats can become deaf in later life, but their other senses allow them to continue functioning fairly normally. Like deaf people, deaf cats learn to 'hear' with their feet, feeling vibrations which we sighted people miss because of our bias towards the senses of sight and hearing.

SMELL AND TASTE

If, by now, you are beginning to imagine the world of the cat as though you have its perception of fairly drab colours but brilliant night vision, its sensitive 'seeing' whiskers and 'hearing' feet, you're ready to be launched into an even stranger world: that of being encompassed by smells so intense that it is like swimming through different colours, textures and tastes of fluids, each giving information about the surroundings – past and present. Our own senses of sight and hearing are fairly good, even in cat terms, but, when it comes to smell, in comparison we are very poorly equipped.

Cats have the same type of smell detector cells as we do in the lining of the nose, with which they identify airborne substances. The membrane on which these olfactory cells are to be found is about the size of a handkerchief which has been folded many, many times. However it is twice the size of the membrane in the human nose and over 200 million cells work to give the cat an amazing sensitivity to smells. Dogs have even more cells, but then they need to track their prey. Cats don't follow their noses when hunting, but stalk their prey using hearing and sight. Their sense of smell comes into play more as a means of communication, to read the messages and marks left by fellow felines and also, usually inadvertently, by other members of their group including humans and other pets.

Taste receptors on the tongue enable the cat to recognise substances which it licks, laps up or chews. The cat's tongue functions as a comb as well as an organ of taste. Down the centre are backward-facing hooks which help to hold prey as well as to lick food or remove tangles from the coat. Cells on the tongue are sensitive to temperature and taste – even to that of water – but, interestingly, cats do not have a sense of 'sweet' (some do seem to have what we would call a sweet tooth, enjoying chocolate and cake, but just what they are tasting is a mystery). Since cats are fundamentally carnivores, it is no surprise to learn that their taste is stimulated by chemicals which are constituents of proteins. Fats in different meats not only taste different but cats are able to tell them apart by sniffing them, so the discerning feline can turn up his nose at chicken or beef and demand the smoked salmon simply by catching a whiff as he walks by.

And if a very good sense of smell and a selective sense of taste are not enough, cats can also use a further method, one which combines taste and smell. When chemicals in the air (which are, after all, what smells consist of) are trapped in the mouth, the cat presses its tongue against the roof of its mouth to transfer them into a thin tube of cartilage, about half an inch long, which is situated above the roof of the mouth and opens up just behind the cat's front teeth. This is called the vomeronasal organ, or Jacobson's organ, which seems to enable the cat to concentrate smells and taste them at the same time. With certain smells, such as those in the urine of females in season, the cat that smells them exhibits a strange behaviour: it stops, stretches its neck, opens its mouth and curls up its top lip. In this way it draws as much air as possible into the opening of the Jacobson's organ so it can be smelled/tasted. The cat can retain this sniffed air in the cavity during normal breathing. This grimace-like response is called the Flehmen (pronounced 'flay-men') reaction. It is also seen in deer and horses, in which, because they are larger and perform it more frequently, being more social animals, it is easier to spot.

Although the Flehmen reaction is more frequently seen in toms sniffing the urine of female cats, it can be observed in both sexes, neutered or entire. I noticed our cat Flirt Flehmen when she sniffed a wet nappy from our new baby for the first time. The response can also be seen in reaction to catnip or catmint. A certain chemical in the plant seems to give about 50 per cent of cats a 'high' and they tread, roll and rub along the ground as in pre-mating behaviour, miaowing and often getting rather excited. It may be that the smell is one similar to that emitted by females in season. It has been suggested that catnip affects the sensitivity of the skin on the chin and head and stimulates the glands there to cause the cat to rub against almost anything nearby. Again, entire toms (unneutered males) tend to react more ecstatically, but many cats seem to enjoy this psychedelic state induced by the plants, and catnip is often put in cat toys for this very reason. The same biochemical pathways are affected by the smell of catnip as by marijuana or LSD in man, although the effect for cats is short-lived, non-addictive and harmless. It is not just domestic cats who react in this way; African lions too will rub, roll and bite the plant.

When a cat is old or ill, especially when suffering from respiratory tract infections, its senses of smell and taste may become blunted and its appetite may dwindle. Offering food heated to body temperature to release odours, or providing strong-smelling meals of smoked fish or liver, may stimulate the taste-buds and help the cat regain its appetite.

FLEHMEN

You have to watch your cat carefully to spot the Flehmen reaction – the cat raises its head and pulls back its lips slightly, drawing air into a special organ in the roof of its mouth to taste/smell the air.

About 50 per cent of cats react in the same way when they smell the catnip plant. Catnip is thought to affect the same chemical pathways as LSD in man, although the effect is short-lived, harmless and non-addictive in cats. It is often put in cats' toys to encourage the rubbing, rolling and mewing reaction it also elicits.

Once we appreciate how very strong is the sense of smell in the cat and what it can tell the cat about the world of other animals and other cats, we are better equipped when it comes to feeding, cleaning and understanding its behaviour.

SIXTH SENSE?

For many years stories of cats with strange abilities to foretell natural disaster or weather changes have abounded. Most reports concern cats showing strange behaviour before storms, eruption of volcanoes, earthquakes or even before less natural phenomena such as air raids. (Many households with a cat in World War II soon realised they could regard puss as a reliable early-warning system as it became agitated before sirens went off.) Often stories involve a mother cat moving her kittens from an area or house which later is devastated by flood, landslide or lava to a safer refuge; or a cat trapped within four walls desperately trying to get out. Reports of such instances of forewarning, as displayed by many cats before the 1979 California earthquake, are now taken so seriously in America that scientists in seismology there are studying the behaviour of over 200 species of animals with the help of 10,000 volunteer observers. If these observers notice anything peculiar in the animals' behaviour, they have to dial a hotline to the earthquake scientists. This is also done in China and in 1975, acting on the behaviour of cats and other animals, Haicheng was evacuated twenty-four hours before a huge earthquake devastated the area. No doubt the cats had already taken to the hills. And, let's face it, if you live in a region likely to suffer from quakes, all forewarnings are gratefully received.

There are several theories as to how cats seem to be able to predict these happenings. During rainstorms, enormous amounts of electricity are discharged into the clouds and create electromagnetic waves that spread through the atmosphere for hundreds of miles. The air becomes charged with positive ions which are thought to influence the concentration of certain chemicals in the brain. As a result some people suffer a headache before thunder. The cat may be many times more sensitive to these ions, and the changes within its brain cause its mood and behaviour to alter dramatically. They may also have the ability to use Jacobson's organ to sample molecules in very dilute concentrations in the air and get a forewarning of more violent changes to come – such as when a volcano has begun to smoke, and so releases the gases within it, but has given no great outward sign of eruption. Some cats are said to rub their ears before heavy rain, responding, perhaps, to pressure changes that agitate the sensitive inner ear.

The sensitivity of the cat's feet and whiskers to vibration may mean that it can sense the tiny tremors which precede an earthquake. If we take into account this awareness of vibrations, the ability of the cat to hear ultrasonic sounds and to detect magnetic changes, then a storm or earthquake may be as obvious to a cat as an air-raid siren is to us, and might well be detectable hours before we humans become aware of it in its much less subtle form.

While prediction of earthquakes or other physical phenomena can perhaps be explained by modern science, there remain several vexed questions concerning a 'sixth sense' in cats. There are many reports of cats that anticipate the return of their owners after a long absence and without obvious warnings. There are also many, many stories, some corroborated by evidence rather than just anecdotal accounts, of cats travelling hundreds of miles after being left behind on holiday or returning to their old home after a move. Cats do appear to have remarkable navigational powers, perhaps thanks to an in-built magnetic sensitivity which gives them the same homing ability as is found in pigeons. They also have an amazingly accurate 'internal clock' and will welcome the kids home from school at the same time each day or wait by the food bowl at the right hour every evening.

Stranger still are tales of cats which have left home to find their owners at great distances away, in places that they have never been to before. One such story is of a cat whose owners were due to move to a new house 200 miles away. On the day of the move the cat somehow got left behind, but it turned up later at the new house! How these cats can even begin to figure out the right direction to take, let alone pinpoint the location of a house so very far away is a complete mystery, but this is by no means a unique story. There are sufficient reports of similar strange happenings to make us want to find out more about just how the cat, or rather, some cats, are able to do it.

TIPS: ENTERING YOUR CAT'S WORLD

Use your imagination to get into your cat's skin, to see through its eyes, and a new world will open up before you. Imagine:

- Feeling texture and vibration each time you put down your bare feet – as though walking on your hands.

- Wanting to get up higher, so leaping up twenty feet without a second thought.

- Falling from the highest diving hoard and, instead of belly-flopping into the water, automatically twisting like an acrobat so your feet hit the water first.

- Being able to bend and twist with a spine so supple that you can touch your back with your tongue and put your leg over your head.

- Knowing your body is fairly insensitive to changes in temperature but finding your nose and lips are as sensitive as human fingertips.

- Seeing the world around is coloured in dull shades of blue and green and blurred at the edges.

- Going out at night where there are no lights at all and not stumbling about in the dark – being able to see as if using one of those infra-red cameras the military use for night manoeuvres.

- Having a 'force field' of hairs around your face which can feel a tiny breeze lighter than a cobweb you brush past in the dark.

- Hearing the tiniest squeak and being able to pinpoint the mouse immediately.

- Entering a room with your eyes shut and being able to tell not only who's there but who has visited recently.

- Tasting smells and being able to identify meat by its odour.

2

Cat Talk

The cat, armed with all its super senses, does not merely use them to catch its daily meals. They allow it to enjoy a rich and complex social life, communicating with other cats and with us using body language, vocalisation and scent. We can, with a little observation, understand some of the cat's body language (although the subtleties are probably lost on us) and interpret the intonations of its calls, miaows and purrs. The third medium, that of scent, is all but lost to us – unless we are 'lucky' enough to get a whiff of good old tomcat spray! But it is the sense of smell which is the most powerful of all the cat's senses and an integral part of every moment of its life.

SMELL TALK

A blind defenceless new-born kitten uses its well-developed senses of smell, touch and warmth detection to guide itself to its mother's nipple. Using scent it comes to know its own particular nipple, to which it will try to return at each feed. The queen knows her young by their smell and by her scent on them and the importance of scent as a mode of communication is established.

EXCHANGING SMELLS

Cats have certain areas of skin on the chin, lips, temples and at the base of the tail where special sebaceous glands produce an oily secretion specific to each cat – its own calling card. They use this scent to mark areas around them, other cats, people and other animals within their group. Stroke your cat and notice how it rubs its chin, lips and head along your hand. Touching cats in these certain areas seems to give them even more pleasure than they enjoy when you merely stroke their heads and backs. By tickling them we further spread their scent and mix it with ours. Lions will rub heads and smear each other with scent to create a collective or communal smell comprising elements of all friendly members of the pride. In this way they recognise each other, as well as instantly being able to detect an intruder. To become accepted within the group, any newcomer must mingle and take on the group identity. He will probably be looked on with some distrust until he has 'earned' the right to belong. During this probationary period he will gradually become anointed with the group smell, which, of course, will itself change slightly to incorporate his own personal scent too. He will then have the correct 'membership card' and be recognised and welcomed by the other cats as we would accept and remember a new face.

Although we may not consciously realise it, humans have retained some of this ability to recognise smell. Experiments with humans have shown that human mothers can pick out T-shirts worn by their children and can recognise the smell of their own baby a few days after birth. Other people recognised a 'family odour' in T-shirts worn by mother and child, and people could detect their own smell or that of a partner on a shirt. So we too have some ability to recognise a family or group smell. In these days of scented antiperspirants and deodorants we do our best to remove all traces of what we call 'BO'. Our own body odour is often offensive to us, yet some of the most expensive perfumes we use contain secretions from the anal sacs of various animals, including civets, cat-like creatures from Africa and Asia.

For the pet cat, its group, its home and its territory have a definite smell profile. A new smell, such as that of a baby, another animal which has visited, or merely a new piece of furniture, may alter this status quo and upset the cat's concord with its environment, although usually only temporarily. However, if the cat is very sensitive and easily unsettled, a disturbing new smell can sometimes cause problems such as indoor spraying of urine. The cat attempts to reassure itself in the face of challenge or upset by adding its own smell to the den to overcome the insecurity brought on by the strange one. Later on in this book you will find some tips on how to recognise and overcome just such problems.

LEAVING SUBTLE MESSAGES

When cats that are familiar and friendly to each other meet, they rub head, flank and tail against one another, exchanging odours and greetings, just as we would shake hands or kiss an acquaintance and make light conversation. Their straight-up tail stance allows them to investigate each other's anal region where glands are situated under and above the tail. You may notice that when you stroke a cat it seems to stand on the toes of its back feet and lifts its rear end into the air as your hand moves along its back. This is actually an effort to raise the gland into the air to give scent and to let you investigate him. When cats wash, they sit with back legs stretched out before them and lick from the anal and genital region outwards, down the thighs and tail, spreading their scent over a wide area that is normally 'pocketed' by the tail as the cat walks along. Grooming also stimulates the glands to produce more of the secretions, as well as spreading them over a wide area. When one cat meets another cat or a human friend, the tail is raised quickly, 'letting out' the smell and encouraging further investigation.

Cats also leave smells around their home range or territory for other cats to find. By rubbing lightly against twigs or leaves they leave a smear of an oil-based secretion containing their scent. The cat may also deliberately rub its chin several times against the top of a stick and raise its top lip, as if sneering, to anoint the tip with secretions from the glands around the mouth. Other cats take a great deal of interest in these 'scent sticks' and may move from one to another, often over-marking them with their own scent as they pass. By investigating each one they can tell when and in what direction other cats passed and can lay down their own message. This not only leaves a message for other cats to find but also gives the message-layer a feeling of confidence.

Small gaps in fences and hedges through which cats squeeze when on patrol also become smeared with body oil and scent, as well as hair, and give similar clues to the cat's passage and occupancy of a home range.

Cats often 'timeshare' their ranges with other cats, and so perhaps when it is their 'turn', they like to have their own familiar smell around them. It can be likened to unpacking all your belongings and spreading them around when you stay in a hotel room – it makes you feel more at home. The old belief that every cat has an exclusive territory within which other cats will fear to tread no longer stands, especially in urban districts where the concentration of cats within a very small area is great. Usually the more assertive cat will have free access to his territory whenever he wants it, this usually being at the important hunting times of dawn and dusk. Other cats will avoid the area at these times but use them, say, at midday when the dominant animal is less likely to chase them away. Resources are available to them on a timeshare basis.

RUBBING

Cats rub themselves around objects in the house and garden as well as people and other animals, not simply to satisfy their need for touch but also to leave subtle smell markers. Glands around the chin and lips and at the base of the tail secrete a personal odour which the cat uses to anoint its territory and its friends, creating a group and area smell.

SCRATCH 'N' SNIFF

Although they have sweat glands all over the body, cats have sweat glands similar to our own only on the pads of their feet. These secrete a watery sweat when the cat is hot or frightened, and it may leave a trail of damp footprints as it walks across the floor. These secretions keep the pads moist and prevent cracking and flaking (as happens on the pads of dogs' paws) so that the pads remain sensitive and flexible. Sensitivity is especially important during night-hunting when the cat needs to be able to feel what it is walking on while keeping its eyes firmly fixed on its prey.

When the cat scratches a tree, door, or even an armchair, it is not only keeping its weapons in trim by removing the old husk from its claws to reveal sharp new points underneath, it is also adding a scent marker to its visible scratch marks. Thus, scratching posts are often used not merely as the means to sharpen claws and stretch; again, they act as flags to reveal the cat's presence in the area. If practised with great vigour in front of other cats, scratching is also a signal of self-confidence, and assertive cats will scratch frequently in front of others as a type of macho display.

SPRAYING AND SCRATCHING

When spraying, cats take up a very characteristic position with their tail held high and quivering. They tread up and down with their back feet and squirt a small volume of urine on to a vertical surface behind. Scratching may not just be a claw-sharpening exercise – glands between the pads ensure that the cat's scent is also smeared over the scratched area.

MESSAGES THAT CAN'T BE MISSED

Rubbing and scratching are the more subtle and intimate forms of communication that a cat uses to mark its presence. In the garden, or anywhere outside the core of what the cat considers its home base, both male and female cats have a system which broadcasts their presence a little further – they spray urine. Both neutered and entire cats may spray, contrary to popular belief that only uncastrated toms do it.

When cats spray urine they take up a very characteristic position with their tail held high and there then follows a quivering action accompanied by a paddling or treading motion with their back feet. A fine spray of about one millilitre of urine is squirted straight backwards, usually on to a prominent vertical surface. While urine, and faeces also, may be used as markers, the impact of a nose-level spatter of urine which, especially if deposited by an unneutered tom who has added other pungent chemicals from his anal glands, cannot be passed by unnoticed. It is thought that on a still day tom-cat urine can be detected by another cat from forty feet away! These signals may last for up to two weeks, depending on the weather, and certainly for days afterwards other cats can tell a great deal about the sprayer. These chemical markers decompose at a fixed rate (if they don't get rained on and washed away), which means that a visiting cat can tell how long ago the sprayer passed by, its age, sexual status and probably even its identity. In this way many cats can live in overlapping territories and use their spray markers like flags flying to show when a particular queen (or tom) is around and active in the territory.

Watch your cat in the garden – you will soon discover its marking spots, see them being sprayed and notice how other cats sniff the message and perhaps add one of their own. Closer observation may reveal the more subtle scent-sticks where cats rub their chin to leave a message. If you live in an area with many feral cats, including different unneutered toms and queens, there will probably be even more enthusiastic marking behaviours and an odour that people can readily detect! The uric acid in the urine is also very corrosive and can affect iron and other metals, rot wood and damage plants.

Tom cats in particular may also leave their faeces and urine uncovered along the borders or at frequently challenged posts of their territory. They often deposit them on tree stumps or in the middle of a walkway where they may be encountered by a rival. Known as middening, this is a very unsubtle marking ploy used by many animals in the wild. Smell is a very useful form of communication when, like the cat, you are active in periods of darkness when visual signs such as scratch marks may be less visible – there's no mistaking a strong smell at any time of day!

We must also realise that we, inadvertently, may also carry in scents to our own cat's territory. Walking through an area with a high concentration of cats (any suburban street, for example) we may pick up on our shoes urine from dogs or other cats and carry it home. Our cars parked away from home are often sprayed in the street because they bring in smells from many areas. When we then drive home, our own cat thinks that a strange new cat has moved into his area.

One such cat began spraying inside its own home, much to the dismay of its owners. They traced the beginning of the smelly habit back to their son's birthday – they had bought him a new bicycle which he protectively wheeled into the hall every night, bringing with it a whole host of new smells on its wheels. These new odours disturbed the cat's smell profile of its home, making it feel insecure. In spraying, it was trying to bolster up its self-confidence in the face of invisible opposition. When the bicycle was found a new safe parking-place in the shed at night, the cat settled down again and the problem stopped because the phantom sprayer was no longer around.

A similar incident concerned a cat living permanently indoors in a high-rise flat. Its owners were able to reduce greatly its spraying behaviour by taking off their shoes when they came in and leaving them, and all the offending smells which clung to them, outside. The cat felt secure again in its own home and its need to spray was removed. Indoor cats obviously have a much more fixed smell profile to their den and they may consider any small change to be a great intrusion. Most cats can deal with new smells quite happily, but these 'mishaps' illustrate just how sensitive cats are to 'smell talk' and how we too must be aware of its impact on them.

BODY TALK

Cats are known for their independence and their solitary hunting techniques. Unlike their canine cousins, they do not co-operate to hunt or group together for protection and therefore don't have the social rules of the pack. However, they do mix with other cats, not only for the purposes of mating or raising kittens, but for what we would call more 'sociable' interaction. They have a complex body language – scientists have noted twenty-five different visual signals used in sixteen combinations; no doubt many of the cat's more subtle nuances pass unnoted. But if we can master the recognition and translation of a cat's basic expressions, then we are well on the way to understanding what our cats are feeling and meaning to 'say'.

Most encounters between strange cats occur outside, and in high-density suburban or urban environments cats will meet many others within a small area. The most dramatic body language occurs during encounters between rival males and the most obvious between cats during courtship. Since most pet cats are neutered and we rarely watch them outdoors, most of us have to be content to observe the less extreme interactions between cats, cats and other animals, and cats and ourselves.

Some of the most appealing aspects of a cat's behavioural repertoire are displayed during play. Kittens, and even adults, will put on for us a playful pantomime which includes hunting, fighting or courtship – in fact, the whole gamut of behaviour. The play-acting can be likened to tribal dances where wars and courtships are re-enacted as a learning, as well as a social, process. In most cases, cats which share a home get on well together and interactions are friendly and calm. Where there is friction, you have a much better chance to see the cat's entire repertoire!

Of course, most feline confrontations do not end in fights – even when they are between rival toms. The aim of language is to put your message across, avoid – or at least not prolong – confrontation and to prevent injury. An aggressor will put on a full display to try to make the other cat get the message and run away without resort to tooth and claw. Quite often cats will simply have a long staring match (with a few vocal insults thrown in), which is sufficient for them to decide who comes out on top. The dominant aggressor may merely walk away from the loser, sit down and look in a different direction or even groom, and human observers will be totally unaware that anything has gone on at all.

In this chapter, body language has initially been divided into head (eyes, ears, whiskers and mouth) and body (tail, position, size and angle). As regards positions, however, since some expressions of fear and anger can be very similar, the signs from individual 'components' may be conflicting and easily misread. To get the whole story, the entire body must be taken into account, as the examples will show. Isolating one feature may also be misleading because signals often change rapidly as the cat's mood and mind alters.

Looking at head position, ears, eyes and whiskers can tell us a lot about what the cat is feeling. The many muscles of the cat's face give it the ability to display a wide range of expressions, while the position of the head itself can give some clue as to whether the cat is trying to invite contact or attempting to become invisible. When the head is stretched forward, the cat is trying to encourage touch or to see another cat's facial expressions or those of its owner – the best example of this is when you come home and your cat wants to greet you and to be fed! If in conflict, an assertive animal may raise its head, but an aggressive one may lower it. An inferior cat may also lower his head but, if fearful and defensively aggressive, may raise it! A lack of interest is indicated by keeping the head down, pulling in the chin and turning sideways to prevent eye contact. Obviously it is very difficult, if not highly confusing, to try to guess the cat's mood by looking only at the position of its head, but by considering its body at the same time we can gain enough clues for an educated guess.

EYES

Eye contact is one of the most important aspects of human communication and in cats too it plays an important role, but we must realise that, in the same way that a tooth-baring chimpanzee is not smiling but is actually showing fear, prolonged eye contact between cats is not a friendly action as it is in humans. Staring is certainly an assertive behaviour among cats, and rivals will try to out-stare each other to resolve conflicts. When a cat realises it is being watched or stared at, often it will stop in mid-groom or wake and look up. Although it may then continue with what it was doing, it does so in what we would call a 'self-conscious' manner until the observer has looked away also. We too can sometimes get that unnerving feeling of 'being watched', but cats are obviously particularly sensitive to observation.

At the other extreme, cats are often described as 'day-dreaming' or gazing into space because they sit quietly, appearing not to be looking at anything in particular. This is because cats take in a great deal of information through the edges of their eyes, whereas we view more through the central portion and so look directly at things around us. Cats tend to use this peripheral vision unless they need to 'fix' their eyes on something, such as a moving target.

EYES

Left: Dilated pupils can be a sign of fear or excitement. Pricked ears indicate that the cat is aroused through interest, not fear.

Centre: A relaxed cat watching the world go by. The pupils dilate or constrict according to the available light and degree of arousal.

Right: Alert and ready for action in the daytime, the cat is cautious but not fearful. At night the pupils will be more dilated.

The eyes are good indicators of mood in cats. A narrowing or widening of the eye can display interest, anger or fear, in the same way as it does in man. The size of the pupil is not only governed by the amount of available light, but by the cat's emotions. Pupils may dilate because of either fear or aggressive excitement and, in less dramatic circumstances, when the cat sees its dinner bowl being filled, or becomes excited or aroused by the sight of a friendly cat.

When a cat is happy and relaxed, the pupils will be as dilated or constricted as the available light demands – the less light there is, the wider the pupil and the blacker the eye seems. A relaxed cat will probably not have its eyes wide open but the eyelids may appear heavy and it may blink slowly as a sign of contentment. Blinking is a reassuring signal between cats and breaks that aggressive stare which makes them feel so uncomfortable.

A fearful cat will often have dilated pupils and, as the fear increases, it may get an almost 'boggle-eyed' look with pupils completely dilated and eyes wide open. On the other hand, an angry cat that is asserting itself, but feeling confident in its position, may have pupils constricted to a slit. Of course there's a huge range of degrees in the dilation of the pupil, depending on the intensity of fear, anger or excitement the cat is feeling, and on the amount of available light, and so it is best for us to look at the eyes in combination with the ears.

EARS

The ears are one of the cat's most important instruments of communication. Between twenty and thirty muscles control their movement and they can swivel through 180 degrees and move independently of each other, around, up and down.

In some of the larger members of the cat family pale markings on the back of otherwise dark ears accentuate their position during offensive or defensive encounters, making their owner's message not only easier to see but leaving little margin for error in translating its message. Some wild cats, such as the bobcat, have a much shorter-than-average tail and so are deprived of one of their methods of communication. To compensate for their lack of tail they have developed tufts on their ears which accentuate their position and give them more opportunity for communication. In our domestic cats the Abyssinian also has little tufts on the tops of its ears.

FACE

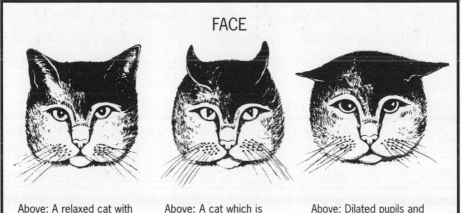

Above: A relaxed cat with perked ears and whiskers. The dilation of the pupil depends on the light.

Above: A cat which is annoyed will turn its ears back, while its pupils constrict and its whiskers bristle forward.

Above: Dilated pupils and flattened ears indicate a fearful cat. Its whiskers may be pulled back.

Left: When playing or hunting, ears are pricked, pupils dilated and whiskers thrust forward.

Right: A contented cat will half-close its eyes, relax its whiskers and doze off in a warm spot.

A happy, relaxed cat will usually sit with its ears facing forward but tilted slightly back. When its attention is caught by a noise or movement, its ears will be pulled more upright and become more 'pricked' as the muscles in the forehead pull them in. It's rather like us wrinkling our brow when we concentrate. If the ears begin to twitch or swivel, the cat is probably feeling anxious or unsure of a noise or situation.

As anxiety increases, the cat moves its ears slightly back and down into a more flattened position. My pair of Siamese cats, Bean and Flirt, exhibit this ear-dipping beautifully when our bumptious puppy approaches their basket where they luxuriate by the wood burner. As he gets within their flight distance (the distance they need between them and him in order to get away successfully), their ears flatten and each takes on the appearance of a dejected gremlin, like the cuddly one in the film of the same name! Their ears end up lying almost parallel to the top of their heads, which they lower into the basket hoping the puppy will then lose sight of them and go away. This is a perfect example of a submissive defence reaction where the cat tries to disappear by making itself as small and unprovoking of attack as possible. When Bullet, our moggie, is faced with the same problem of trying to rid himself of the pup's attention, he uses the more 'active' approach of attacking before he is attacked. He sets himself in a position where he can corner the pup and then frighten him with a startling hiss, arch his back, fluff up his tail so he is noticed and noted as 'large', and start a battle of wits with a good stare. His tensed ears swivel so that the inside surfaces are directed to the side. The backs of his ears are then visible from the front. His hackles go up and he waits until the pup bounces into range before having a swipe at his nose. No one ever gets hurt and both of them seem to enjoy the mock battle. Bullet always wins, of course!

WHISKERS

Above: When a cat is relaxed or indifferent to what is going on around it its whiskers stand out on either side of its face. However, watch a cat yawn and you will see the range of movement the whiskers are actually capable of, from lying flat against its face to fanning out and round in front of its nose and mouth. A cat on the defensive, as shown here, will hold its whiskers back, close to its face. Its ears will also be held back and down and its pupils may be dilated in fear.

Above: An inquisitive cat may fan its whiskers forward using them almost like a third hand (or paw) to feel what is immediately ahead of it. Whiskers are used to see/feel in the dark when the cat moves through the undergrowth or negotiates small gaps.

WHISKERS

Whiskers too can indicate mood and are much more mobile than we realise. When relaxed they are held slightly to the side but as the cat's interest increases they come forward in front of the muzzle. A fearful cat will pull its whiskers right back alongside its cheeks – making its face as small and unthreatening as possible in the hope of avoiding conflict.

The whiskers accentuate the muscle movements around the mouth and in fat-faced toms (in neutered cats removal of the primary male hormones means that the fuller face does not develop) close observation may reveal cheek muscles pulling down the cheek ruffs when the cat is excited or fearful. Wild cats have fuller faces or extra hair, as in the lion's mane.

MOUTH

The cat does not use its mouth in aggressive confrontation as the dog does – a cat's open-mouthed hiss and snarl is brought on by a feeling of threat. Licking of the lips may be a sign of anxiety, although sitting with the tongue hanging or sticking out seems to be a sign of relaxation or contentment, often giving the cat that amusing 'simple' look.

When humans yawn for reasons of tiredness or boredom they can start off a chain reaction of yawning that is almost impossible to stifle in whoever sees them. Yawning is not contagious in cats, nor is it a sign of boredom; it is more a signal of reassurance and contentment and often accompanies a languid stretch after they have awoken from sleep.

TAIL

The tail is used for communication as well as balance. When hunting, the cat holds its tail in a streamlined manner behind until it is required as balance for the final rush. It may also announce the cat's interest and concentration with a twitching movement as it corners its prey. However, the tail comes to the fore as a tool of communication when the cat interacts with other cats or with people. It has a whole range of movements, from side to side, and up and down, in speeds ranging from a graceful slow sweep to a thrashing whip. The tail can change from a sleek coil folded around the cat when it is asleep to an erect bristling brush when the cat is frightened.

A relaxed, confident and alert cat walking through its territory will merely let its tail follow behind until it meets another cat or finds a point of interest such as a spraying or scratching post. If he or she decides to spray, then the typical tail-up posture and foot-treading motion quickly follow as the post is anointed with urine. When it sees a known friendly cat or its owner, the cat will often quickly flick its tail into a vertical position, pulled slightly forward over its back and kinked down a little at the tip. This position allows a friend – cat, or person (if they so wish!) – to investigate the exposed region under the tail where a recognisable scent will confirm that the cat is a part of the group. The greeting is often accompanied by a murmur or chirrup. Kittens greeting their mother will run up with their tails in the air and proceed to let them drop over their mother's rump and rub over the top of her tail in an attempt to solicit some of the food she has brought to the den. Adult cats will do the same with their human owner, rubbing and wiping their tails around legs, hands or even plates in the hope of being fed or fussed over.

TAIL

Above: Typical greeting posture to a friend. The tail is held high with the tip slightly bent forward.

Above: When walking the tail is held relaxed behind the cat at a slightly downward angle.

When not being used for communication the tail is used for balance during jumping and climbing.

Left: The best time to see an 'inverted-u' shaped tail is when cats or kittens are at play – usually during that mad half hour when they chase each other around the house. The cats seem to enjoy that mixture of high excitement and fear which children also love. The hair on the tail may also be fluffed up like a bottle-brush.

Cats at rest, but preparing for action, may watch what they are about to get involved with and sweep their tails haphazardly from side to side. It's almost as if they're mulling over the idea and deciding whether to go ahead or not. As the cat becomes more alert, the tail may swish faster and in a wider arc. This can be the first sign of anger, or it may be intended simply to tantalise or encourage another cat to join in a bout of play. Violent thrashing indicates high excitement or imminent aggression – a signal not easily missed. A wagging tail in a cat means the complete opposite to that of the cheerful, friendly dog: it is a sign that the cat is in some state of emotional conflict.

If a cat feels itself to be seriously threatened by an attacker, it may become so defensive that the hairs on its tail become erect and it bristles like a bottle-brush. Held straight in the air, it can be fluffed up to at least double its size and this pilot-erection is used for display when the animal is defending itself and trying to look as large as possible. Cats in conflict may also hold their tails at a strange angle like an inverted capital 'L' – that is, the tail continues the line of the back before bending and pointing straight down to the ground. This can also be seen in kittens at play, as can the inverted 'U' tail. When kittens or cats play chase or when they have that explosion of energy during a 'mad half hour' when they rush around the house bounding over anything in their paths, be it people, dogs or chairs, they often hold their tails in this upside-down horseshoe shape which, when all its hair is erected, makes the madcaps look as if they have 'the wind under their tail'.

POSTURES

Making yourself look bigger than you actually are is a trick employed by many animals for defence and attack. Often conflict is successfully avoided and the rival backs off when the bluff is convincing enough.

Cats too use this ploy and also straighten their legs to look taller if on the attack. As the cat's hind legs are longer than the forelegs, its body slopes down towards the head from the higher rear end. An aggressive attacking cat will erect the hair along its spine and tail like a ridge, again making it look more impressive.

By contrast, a very frightened cat that is cornered or desperately trying to deter an attacker will erect the hair not just along its back but over its whole body. It positions itself side-on to its aggressor, with its back arched to make itself appear as big as possible, in the hope that its attacker will think twice about proceeding against a larger foe. Having stalled the attack the potential victim cat may then move sideways with a crab-like motion towards a safer area, keeping its eye on the foe in case even this slow retreat stimulates a sudden attack.

A submissive frightened cat may shrink into a crouch and try to look as small and unthreatening as possible in order not to provoke attack or draw attention. It is the opposite posture to a big defensive display and is often practised by a less confident cat that simply wants to be left alone to live a quiet life. It may then roll over on to its back with its head turned to face its attacker as an ameliorative, appeasing gesture. This is not quite the same as the submissive belly-up posture in the dog, which, from such a position, is very unlikely to fight back. The appeasing cat has been pushed as far as it will go in its attempt to avoid conflict and is as ready as it can be with teeth and claws to defend itself to the end if provoked or attacked further.

Of course, cats roll over in play, and sometimes also when they greet us. Many don't at all mind having their stomachs tickled, and some clearly enjoy it. Females in season will roll in front of males to solicit their attention. In summer, cats enjoy a dust or sand bath as much as the birds do, and they roll over and over on a warm sunny path, usually keeping a keen look-out in case a passing cat or dog catches them in that most vulnerable of positions.

POSTURES

Above: During hunting and stalking the tail can often indicate interest by wagging slightly at the end.

During the rush and pounce the tail is used to balance the body and help ensure an accurate strike.

Left: The amplitude of horizontal swing of the tail can indicate the degree of agitation the cat is feeling. From a gentle twitch at the tip denoting interest, it can build to a wide fast sweep when the cat is highly aroused, perhaps at seeing a rival in its garden.

Above: As the cat grows more defensive the hair on its body bristles even more and its tail is held arched over its back.

Right: The 'Inverted-L' tail is often seen in a cat which is the aggressor in a conflict. Its back is slightly arched and its fur bristles along its spine and on its tail. The front paw is often raised.

Above: A defensive cat will stand with its back arched and its body at an angle to its aggressor. The hair on its body and tail bristles.

TALK TALK

'And what does the cat say?' we ask our small children. 'Miaow,' they reply. 'And what does the doggy say?' 'Woof woof.' While 'woof' may be a fairly accurate representation of much of the dog's attention-demanding verbal communication with us, a simple miaow falls far short of being representative of the huge variation of sounds the cat is capable of, and greatly understates the range of sounds it uses with us.

Not much research time has been spent trying to fathom the intricacies of the feline vocabulary and those who have studied the subject have categorised the sounds in different ways, but it is probably best to look at them in the context of our interactions with our pet.

Listening to the many different sounds and intonations a cat employs in communicating with us requires more than a little concentration. We usually react instinctively when the cat 'talks' to us, not listening to it as much as looking for visual clues as to the source of the cat's desires – does it want a cuddle, its dinner or just a reassuring chat? The clever cat makes it easier for us by going to the empty food bowl or to the door, or paws the latch on the window if it wants to be let out. If you are really keen to differentiate between the sounds your cat makes, tape them using a portable recorder and then talk on yourself about what the cat seems to want. Listening to the recording later, without the benefit of the cat's body language and movement clues, may enable you to identify the more common feline expressions, calls or requests or the particular ones your cat uses with you.

Cats do have their own individual repertoires of sounds and actions. Not only that, they are probably much better attuned to ours than we are to theirs. They adapt what they say to how we respond, almost as if they have trained us to react to certain sounds. If we respond in a positive way to a certain mew, they are likely to use it again and the resulting success ensures that their vocal repertoire for use with people, or perhaps one individual in the family, expands as we cotton on. Often they have 'words' for speaking to us that they do not use with other cats or even with other people in our family.

The different breeds not only have varying leanings towards vocal conversation (Siamese cats and other Orientals are known for their chatty natures), they place a different emphasis on the sounds or pronounce them differently (as with human regional accents perhaps) and we can often tell our own cat's voice from others'. They may also make additional characteristic sounds identifiable with the particular breed, such as certain chirrup or clicking sounds, just as different breeds of dog sound different when they bark. When Bean, one of our Siamese kittens, disappeared I combed the area calling her name until, at about six o'clock in the morning, I heard a cry emanating from a neighbour's shed. There was no mistaking her voice coming from the far corner under an old table. She replied with increasing vigour and desperation every time I called, allowing me to pinpoint her exact location. Her rescue was due entirely to the fact that she is a talkative little soul.

About sixteen different vocalisations have been identified in cats, though no doubt there are many more which are either too subtle for us to differentiate from others, or are in the 'ultrasonic' range. These latter sounds are ones that are beyond our hearing abilities but are certainly not 'ultra' to the cat, which is sensitive to much higher frequencies of sound.

When kittens are very small, their mother uses only a small range of vocal sounds with them. By regularly going into the kittens' nursery we can learn more about cat communication while it is kept simple. The queen uses certain sounds to express distress, greeting or danger and each kitten has its own distress call to gain her attention. By the time they are twelve weeks old the kittens will have mastered the entire repertoire and, unlike human babies, who must hear and repeat sounds in order to learn our language, even a deaf kitten will use all the available vocalisations of cat conversation.

Cats can vocalise and breathe in and out at the same time. Hence they can produce sounds in a slightly different manner from ours. The cat's tongue plays a less important role in forming the different sounds, which are made further back in the throat by pushing air at different speeds over the vocal cords stretched across the voice box. Shifts in the phonetic quality of the noises are achieved by changing the tension in the throat and mouth muscles. Much of the vocalisation is for short-range communication, except of course with those most blatant of all cat sounds – calling and caterwauling during courtship or the warning angry cries of rival toms squaring up over territory disputes.

PURRING

The purr could be said to epitomise the cat and it is one of the reasons we enjoy our feline companions so much – they tell us very audibly when they are contented. It's a tonic to our emotions to hear our cats purr – like being regularly told that we are loved by our partners or family.

All cats purr at the same frequency – 25 cycles per second – irrespective of age, sex or breed.

They begin to purr when they are tiny kittens, and because purring does not interfere with nursing it can continue as a reassurance between mother and kitten that all is well. The mother may also purr as she enters the nest to let the kittens know that she is home and there is no danger. Older kittens will also keep their little engines going if trying to get adult cats to play – probably like a child trying to ingratiate itself while begging dad to 'please, please play football with me'.

More dominant kittens also purr when they want to initiate play with a more submissive litter mate – a case of 'don't run away; trust me, I'm friendly, and I'm not going to beat you up this time . . . honest!' Adult cats also purr to one another in the same reassuring way, and they may also use purring for self-reassurance when in pain or to placate any potential nearby aggressor. This is because purring is associated with a feeling of well-being and the cat may be using it to back itself up when in pain or fear – like you might, in a crisis, chat to yourself or sing in order to keep your morale high and to distract yourself from the dangerous reality of the situation.

Just how a purr is actually produced in the body is still much of a mystery and theories abound. One suggests it is created by a vibrating of the false vocal cords which are positioned next to the vocal cords.

Another holds that turbulence in the bloodstream sets up a vibration in the chest and windpipe and resonates in the sinus cavities of the skull to produce the noise we call a purr. A third theory suggests it is caused by out-of-phase contraction in the muscles of the larynx and diaphragm.

But however it is produced, like the tail-wagging of the dog, which has a similar meaning, purring can vary in intensity and enthusiasm. Cats can keep up their purring rhythm for hours, the sound varying in loudness from the rough purr with a distinctive 'beat' to the smooth, drowsy or almost bored purr with indistinguishable beats that suggest it will probably draw to an end relatively soon. A higher-pitched purr is also often used when the cat is eagerly seeking attention or has spotted something it thinks might give it pleasure.

SOUNDS OF WELCOME

The whole greeting behaviour of the cat – tail up, striding forward watching your face, rubbing and purring – is crowned by its call of welcome. It is a special short mew or a sequence of chirrups which lets you know just how pleased the cat is to see you. Each cat may have a special 'welcome noise' and may even vary it with members of its household, some of whom may receive no welcome at all if they are not favoured! Coming home is a great opportunity to make the time and effort to talk to your cat. It is excited to see you and wants to interact – so make the best of it. If you do not respond it may gradually become less enthusiastic about its interactions with you too. Nothing is nicer than to have an independent loving animal choose to rush up to you and want to 'talk'. Use it as one of those times to unwind. Ask the cat how his day was and your warm tones will encourage him to tell you just what he's been up to.

The chirrup that sounds like a rolled miaow is often used by a queen calling her kittens or a cat making a friendly approach to another cat or person. Bean and Flirty make a certain '*rrr*' sound when Bullet comes in through the cat-flap and they rush to greet him. Cats also have an acknowledgement murmur – a short inhaled purring tone which drops in pitch – as well as a coaxing murmur that they use to encourage the owner to give them something they want.

Professor Paul Leyhausen, who has studied cat behaviour intensively for many years, noticed the various 'gargling'-type sounds a mother cat made when she brought prey for her kittens. She seemed to make one noise which signified a mouse and another, more of a cry, when bringing in a rat. He questions whether the cat was in effect using 'words' which corresponded to the types of prey, so that the kittens knew what she was bringing in. Although an injured mouse could be approached without risk, a rat might well be more dangerous to a small kitten, and indeed the kittens did show much more caution on the utterance of the mother's 'rat sound'. Leyhausen suggests that a cat's ability to convey meaning and our ability to interpret it have been underestimated and that the subject merits much more study.

THE MEANING OF 'MIAOW'

The miaow, mew, or meow has many variations and is uttered with an open mouth which closes on the 'ow' to make what seems to us like a distinct word. Miaows of many types are used to request us to get the food out, open doors, give attention; they are also used to beg and demand, or even to complain.

If we break the 'word' miaow into syllables – '*me-ah-oo-ow*' – it is apparent that the cat can vary the length of each component or emphasise one or more to make different sounds with different meanings. If the '*ah*' component is not stressed the cat sounds 'pathetic' or disappointed. If it lengthens the '*ow*' element too, the situation sounds absolutely hopeless. We often hear these plea sounds when the cat is shut out of a room it wants to be in, or when the Sunday chicken is on the table and it has not been offered any. If you look as though you might be giving in, the sound becomes 'happier' and lighter and may be interspersed with purring to encourage you not to change your mind – it's called manipulation!

When plea turns to beg (in a dignified cat manner, of course) many cats repeat the '*ow*' at the end, closing the mouth slowly and drawing out the message. Adding emphasis can turn the plea into a demand, often used when owners are a little too slow in filling the feeding bowl. Often the short, higher-pitched mews emphasise how desperate the cat is to put its point across. Sometimes it sounds as though the cat is merely using the middle '*ah*' sound – a favourite tactic of my two Siamese when they are desperate to come into the warm after venturing outside on a frosty morning. The cat uses the highly pitched noises – the ones its mother acted on so promptly when it was a kitten – to gain our attention. In fact the first miaow the cat uttered was probably when it wandered from the nest and had to call for help.
Cats often approach and get our attention with a tap of the paw while issuing a 'silent miaow'! It has been suggested they use this most persuasive of pleas with those who don't usually give in to their needs, to a dominant individual in the house, when asking for attention or food. It's a good ploy and it usually works since that tapping paw and half-heard whisper are hard to resist. Of course, the sound is not 'silent' to other cats but it's too high for our less sensitive ears really to detect. Bullet uses the same ploy with me when he wants a lap to climb on to in the evening, so perhaps it is more of an intimate plea used with certain people who have a close relationship with him.

HISSING, SPITTING AND GROWLING
Hissing, spitting and growling are used by the cat to warn or threaten, noises we seldom hear in our indoor relationship with our pet.

Hissing is used as a warning and is produced by the cat opening its mouth, drawing back the upper lip so that the face is wrinkled, and arching the tongue to expel air extremely fast. If you are close enough you can feel the stream of air rushing past. This is probably the reason why, unlike horses, cats don't like humans to hiss at them by blowing up their noses or on their faces. Producing a hiss affects the sight, hearing and, if close enough, the sense of touch, and the message behind the hiss is very effective. A hiss can be used effectively to let cats know when they are doing something you don't like. A quick 'sssss' as the cat goes to leap up on to the work surface, jump on to the cooker or scratch the furniture usually gets the message across in more than adequate 'felinese' without your having to resort to shouting or physical intervention.

Cats spit both voluntarily at an approaching opponent or almost involuntarily if they get a sudden shock. The spit is usually sudden and violent and often accompanied by both forepaws hitting the ground with a thud in the same way that rabbits thump the ground with their back feet. This is normally a bluff to surprise the opponent into stepping back and allow the spitter to make a break for freedom.

Growling is usually a more aggressive act which can progress to a raised-lip snarling if the perpetrator is attacking a rival. My cats do it quietly if one of them has managed to purloin a tasty bit of chicken and the other two have come to investigate and try to grab it. The low guttural growl from the pilferer usually convinces the other cats that their normally friendly companion actually means it this time and while possession is nine-tenths of the law, the snarl is the remaining one-tenth that deters any nagging doubts in the challengers.

CALLS AND CRIES

Many of the higher-strained noises are reserved by cats for other cats. These sounds go higher than we can hear and have effects on the feline ear and brain that we can't imagine. Within this repertoire some sounds are made with the mouth open, as with the tom-cat caterwaul or the anger wail aimed at another cat on the territory.

Another dramatic cry is the screech of distress, which can also be made by females after mating and is an accentuation of the last syllable of 'miaow' – the cartoon character Tom of *Tom and Jerry* tends to use this one quite a lot! At the moment of ejaculation the female releases a loud piercing cry followed by an almost explosive separation and she turns on the male. It is thought to be caused by the stimulus of the male's penis which is covered with spines at its tip, and the sensation may be painful – it certainly sounds so.

Trying to catch the cat to place it in its basket – a sure sign, as far as the cat is concerned, that it is heading for the vet or a cattery – is often difficult and may cause a cry of protest or refusal aimed directly at the heart of the owner. It is upsetting to hear and you need to grit your teeth and adopt the attitude 'it's for your own good!' to stick to your purpose. Keep chatting and reassuring the cat that everything is fine – it will help to convince you as much as him. Things usually get even more heart-rending when it's time for the cat's annual vaccination at the veterinary surgery. There cats usually take one of two actions: either they shrink into the submissive crouched posture and utter little mews of protest or they become cats out of hell, striking anybody within reach. If your cat opts for the former tactic, you can be helpful and reassuring and even hold him firmly and comfortably while the vet gives the injection or does the examination, talking quietly all the time and instilling confidence with your calm presence (no matter if you are only outwardly calm). If your cat is panic-stricken and resorts to aggression then he will probably not be reassured by anything and it is best to let the veterinary nurse hold him in a safe firm grip which may not look very comfortable for the cat but does ensure that the whole procedure is over and done with quickly and with minimum fuss. Once he's safely back in his basket you can talk to him, and when you arrive home spoil him a little until his indignation passes.

TEETH CHATTERING

Teeth chattering is not really a communication noise. It is often produced when a cat sees something it wants but can't get to, such as a fly high on the wall or a bird on the other side of the window. Perhaps the cat is voicing its frustration by chattering its teeth, like we might shout through gritted teeth. With its mouth slightly open, the lips are pulled back and the jaws open and close rapidly. The noise is a combination of lip smacking and teeth chattering as the cat gets more and more excited. The cat may also emit small bleating/nickering type noises as its teeth chatter.

Our understanding of feline vocal communication could be said to be still in its infancy. The more we study the cat the more we find we don't know. Not only are we making guesses about the sounds we can hear, we have very little idea how much the cat uses its ability to hear in frequencies higher than we can sense in order to make its feelings clear.

TIPS: TALKING AND LISTENING TO YOUR CAT

To understand and join in cat talk, be aware that:

- Strong perfumes affect the cat's sense of smell like a bright white light cuts out our ability to see much around us.

- New carpets or furniture smell as dramatic to the cat as they look to us and may give the cat a desire to bring things back to 'normal' by spraying its own smell over them.

- Staring is an aggressive act among cats. Blinking is a reassuring action.

- Dilated pupils are a sign of fear, so move quietly and slowly.

- Our voices probably seem very low because they only register in the bottom portion of the cat's hearing range. Use a high voice to get the cat's attention.

- Join in with scent-exchanging by tickling under the cat's chin, cheeks and the base of its tail where it meets the back.

- Use a tape recorder to learn cat talk.

- Be willing to be trained by your cat's talk.

- Twitching ears and a swishing tail are signs of anxiety or anger.

3

Living With Your Cat

When cats live with us they can really get under our skin and they seem to make it their business to have their nose or a paw in everything that goes on. They appear from nowhere to watch over the DIY; they lie on the clean ironing or sit in the middle of a newspaper when we are reading; they scramble the wool or string we're using, or merely sit on our laps when we're trying to write a letter. Over the years we come to be at ease with our cats — we learn their likes, dislikes, and habits, and they learn ours.

We can gain even more insight into their behaviour by understanding some of the motivations behind everyday happenings and habits. For example, why is touch so important to cats; how should a cat be greeted or a new cat approached; is there a best place to position a cat-litter tray; how much sleep does a cat need, or just prefer?

MEETING, GREETING AND TALKING

The key's in the lock after a long day at the office. As the door swings open the cat miaows and runs forward, its tail up in that typical greeting posture. As the cat hops with its front feet in mid-air and rubs against your shins, the tension slips away – you're home. Guided by its internal clock, the cat has probably been waiting for the sound of your particular car and your known step on the path when you arrive.

Chirruping and purring, he'll weave between your legs, rubbing his head and flanks against you, anointing you with his smell and the essence of home. The typical greeting behaviour between cats that know and like each other is to walk forward, tail up and slightly bent forward at the tip – ready to meet face to face, to sniff and touch noses, rub heads and lick ears and then to sniff under the tail. Some cats bow their heads slightly before rubbing heads, and they may do this with us too, adding other purrs and postures designed just for 'their people' and stretching upwards to be tickled and stroked. They may also throw themselves on to the floor and roll over.

On coming home to our cat, we chat as we would to another human. For certain the house does not have an empty feel – this small body with the big personality makes sure of that. Often the cat will be shouting for its supper, running to and fro between you and the food cupboard until the hint is taken and, in its impatience, standing on its hind legs to scrape at the door with its front paws. As we chat away, the cat hears our warm tones but probably only recognises the odd word such as 'dinner' or its name, but is probably trying to encourage us to prepare its dinner as much as we are encouraging a reaction to our conversation.

A cat soon learns its name, usually faster if it is a word of one syllable. Sooty has been the most common name for cats in the UK in several surveys carried out on pet names, Other classics are Kitty, Pussy, Tiggy, Tiger, Tigs, Tigger or Fluffy. Alcoholic drinks figure highly – Brandy, Shandy, Sherry, Pernod and Whisky are frequently used. Colour descriptions too are favourite names – Smokey, Blackie, Snowy, Ginger and, of course, Sooty – whether all the Sootys are black is not known, but some owners call their black cats Snowy just to be different! Garfield and Thomas figure highly and understandably, although Marmaduke and Snoopy are common too, even though they are taken from dog characters. Pairs of cats are often given 'half' a name each – Salt and Pepper, Bubble and Squeak, Cagney and Lacey, Mork and Mindy, Pinky and Perky and even Harley and Davidson – which is all very well until one is lost before the other. Obviously pedigree or show names are long and elaborate but these cats also have pet names to make life easier for everyone concerned. Zaphod Beeblebrox is one of my favourite cat names, as is Baldrick. One person obviously knew her place in life when she called her cat Aysha – meaning 'one who must be obeyed'!

The way we say a cat's name can govern how it responds. Bean will respond, especially vocally, if her name is said in a high-pitched voice. All the family has taken to squeaking to get the reward of a reply from her. Cats are very sensitive to highly pitched sounds from early days when kittens make high mews and calls to their mother. Small rodents, the principal prey of the cat, squeak right at the top of the feline hearing range, so cats have evolved to tune into hear and respond. It's interesting that human male friends trying to elicit a response from Bean are often disappointed because they can't reach just the right pitch. Perhaps it's the laughter from the sidelines which puts her off too, or maybe it appeals to her sense of humour to force them to strike their highest note!

Calling the cat in a higher voice can cut through the lower-pitched background noises that the cat often blanks out. It always remains alert for squeaks and clicks, so perhaps our calls hit the right mark. We can do the same thing using the high note produced by sucking in through pursed lips – like sucking without a straw or making a 'pishwish' call, the hissing part of which also seems to get the cat's attention.

Communicating with our own cats to help them feel confident and secure is easy, but how should a cat that's strange to us or slightly nervous be encouraged? Paws are often used as weapons if cat meetings are not friendly, so perhaps it is best to keep hands out of the way initially. Because friendly cats greet head to head, it is best to get down to the cat's level to begin with and also to approach face to face. Watch the cat but don't stare – this is a threatening behaviour between cats, often a tactic for testing strength and will-power in rivals. Half close your eyes and blink as this will reassure the cat. Twitching of the ears is a sign of anxiety in the cat, so slow down and relax if you see signs of fear. Let the cat come forward and sniff and then introduce your hand slowly at cat shoulder height, not like a bolt from above. The cat will probably stretch his face forward a little once he is sure that your intentions are friendly. Then rub his head and chin to interact more, and maybe even make a purring sound yourself. Confident people-loving cats soon get into the swing of a good fuss and tickle with everyone; others need more time and patience and some may never really relax with anyone but their owners. Some of the reasons and factors which influence how and why cats respond to us are outlined in chapter 5.

Just why cats always seem to go for the lap of the one non-cat-lover in a room is a mystery. Some cat experts suggest that it's because people nervous of cats do not stare at them but sit still, blink and look away, hoping not to attract attention. The cat may see this as a sign of friendliness. Conversely, the cat may be deterred by a cat lover approaching too quickly or trying to entice it into a willing lap by calling and staring to attract attention. Perhaps that person who doesn't want the cat near is also inadvertently giving off certain vibrations or smells that are actually attractive to the cat.

There are certainly 'cat people' to whom cats, even those nervous of mankind in general, will be attracted and whom they approach with confidence and friendliness, often to their owner's surprise and delight.

These people often do not need to make friendly overtures or may not even own cats themselves: they merely have to sit back and let the cats find them. Whether these certain types are the nicest people we know needs a little more study, but why not watch your cat's response and see if he or she is a good judge of character. Certainly I'm always wary of people my normally very friendly bullmastiff dog seems to take a dislike to, but I feel warmly disposed to those my cats select to be cuddled by and crooned over. Is a cat person also a person person?

SLEEP AND CATNAPS

Cats are the world's best sleepers, slumbering away 60 per cent of their lives, which means they spend twice as long asleep as most other mammals. A typical day encompasses over fifteen hours of sleeping and dozing, almost six hours washing and playing, while hunting, eating and exploring make up the rest of the day. Lions will sleep a great deal after gulping down a carcase because the food will keep them ticking over for several days. While herbivores have to munch away all day on vegetation in order to meet their energy requirements, a feed of meat is rich in calories and nutrients. The carnivore must exert more energy to catch its meals, but it is also able to rest and digest between them. Domestic cats, being well fed by their owners, also have spare time in which they sleep. Perhaps this investment in rest goes some way to explaining their longevity when compared with other larger mammals like the dog. Of course, if hungry or cold, or during courtship and mating, cats are more active. New-born kittens sleep 90 per cent of the time but this has reduced to adult levels by the time they are four weeks old. Old cats, like old people, sleep or snooze more often. Cats, like us, also slumber more if warm, secure and well-fed. They often fall in with the daytime pattern of their owners, choosing to sleep when they are alone during the day and being active in the morning and evening when we are around. At weekends they return to regular periods of short catnaps when they take forty winks, and have periods of deeper sleep when they are safe in the knowledge that we are at hand for friendship and comfort if required.

Cats, being superior sleepers, have been used in several scientific studies of sleep. Their brains show similar electrical activity to humans' during sleep. During the first fifteen to twenty minutes cats remain fairly tense around the neck and head and are instantly awakened by clicks, squeaks or sudden noises. They then relax and may roll on to their sides for six or seven minutes, and their whiskers, tail and paws may twitch, perhaps in dreaming. Cats do exhibit what we call REM (rapid eye movement), a characteristic of deep sleep in humans. They may return to shallow sleep for a short period before 'dreaming' again. This deep sleep takes up about 15 per cent of their lives, while shallow sleep accounts for about 45 per cent. Kittens do not shallow sleep in their first month of life because the centre in the brain which controls this lighter sleep does not develop fully until they are about five weeks old.

When 'catnapping', cats will settle on any spot and close their eyes while remaining fairly alert. However, before settling into deeper sleep they need to find a place where they feel secure. After all, this is when they 'switch off' entirely and so they need to be relaxed and safe. (Mind you, we've all had cats which slip from nap to slumber on the back of the sofa and then fall off!) Because the temperature of the body drops during sleep, cats often seek a warm spot in the sun or in the airing cupboard so they feel warm as well as secure.

A napping cat will be well aware that you are around or that you are approaching because his ears will still be on 'radar patrol', scanning for any small sound. The cat that is sound asleep may get quite a shock if awakened by a loud noise or sudden prod, so if you need to waken him treat him as you would wish to be woken yourself – with a gentle whisper, a soft touch and reassurance that everything is OK – and hope that he wakes from deep sleep in a better mood than you do!

Cats are often used in advertising to symbolise warmth, security and comfort within a loving family and are pictured curled up and cosy in front of a roaring fire. One German animal expert studied more than 400 sleeping cats and concluded that he could tell the temperature of the room by the position the cats took up when sleeping. At less than 55°F the cats curled up with their heads tightly tucked into the body but as the temperature increased, the cats' body shape opened up. At over 70°F, the cats were uncurled with paws out in front. Cats may end up with their feet in the air or they may lie flat out on their sides if they feel both safe and very warm. Each cat may have its own particular pattern, which its owner will recognise.

Bear in mind the cat's need for security and warmth when you choose a site for his bed because he'll choose his own spot if not satisfied. Height gives security, especially if there are curious young children or puppies around. Cats climb up to where they can safely watch the world and strike out with a paw should danger come from below. They need safe niches or bolt-holes for a peaceful snooze, which is why many choose a high shelf in an airing cupboard or wardrobe. Don't forget, however suitable and cosy a basket you offer, your cat will use several napping-and-resting places, so don't go overboard on an expensive bed. A cardboard box with an old cushion inside or a blanket in a corner of the sofa is ideal.

On waking, cats usually take a moment or two to stretch and restore their circulation to all parts of the body. With an amazing suppleness that would make even a yoga expert green with envy, they stretch their joints and muscles from top to toe, often digging their claws into the carpet for extra anchorage, arching the body and raising their bottoms high into the air to stretch the hind legs and tail. A few yawns later, a quick face-wash . . . and they are ready for action.

THE JOY OF TOUCH

Touch is very pleasurable for many cats and their pleasure at being groomed by another cat or being stroked by us is obvious. Rubbing enthusiastically and purring in a strong rhythm, they ooze enjoyment. Touch is thought to stimulate the release of chemicals in the brain called endorphines, which give a feeling of pleasure or well-being or even help overcome pain. The cat is thought to have a very active endorphine system and many vets have commented that when some obviously seriously injured cats are brought in for treatment they do not behave as if in pain. Perhaps this is the reason many manage to drag themselves away after road accidents to somewhere quiet: the pain-killing effect allows them to move away to safety without causing further hurt. If stroking also releases these strong psycho-active chemicals, no wonder they purr.

Touch is the primal source of affection and it is known that a human child, monkey or pup may not thrive, and may even die, in the absence of reassuring and loving touch. The same is undoubtedly true for the kitten; indeed the very survival of a new-born kitten is dependent on its mother's touch. Without her stimulating licks on the stomach and under the tail, a kitten is unable to open its bladder or bowel. She licks and cleans the kittens regularly, eating the waste to ensure that the bedding area does not become wet, cold and a breeding ground for infection, all of which conditions could reduce the survival chances of the kittens.

Grooming and touch are vital reassurances to the new-born kitten as it cannot see the world or hear very well. Enveloped in its mother's nest, it is dependent on touch and smell to survive. A cat deprived of touch may be withdrawn and fearful and may even groom itself more in an attempt to compensate for this lack of touch from others. Kittens respond to this safety and comfort by purring – a behaviour they maintain in adulthood in their relationship with us, along with other kitten-like behaviours, in the safety and security of our homes.

Another juvenile behaviour which is often continued is kneading. Cats usually purr very enthusiastically as they are stroked on our laps, almost in rhythm with the paddling with their feet as they draw their claws in and out. This behaviour stems back to the kitten's suckling period when kneading on their mother's stomach around the nipples stimulated milk flow. Even though many years old, cats may revert to this behaviour with us, and we should be flattered that they obviously feel sufficiently secure and content to relax and become defenceless kittens on our laps.

GROOMING

Kittens learn to groom themselves, using their barbed tongues like combs, at about three weeks, often before they can walk. It is an instinctive behaviour at which they are proficient by six weeks of age. They need to care for their coats in order to remove loose hair, stimulate new hair growth, prevent matting and to spread secretions from glands on the skin so that the coat is kept waterproof and their bodies insulated. However, the act of grooming is itself also a way of keeping cool since saliva evaporates off the coat and removes excess heat. Hence cats groom more in warm weather or after strenuous activity and may lose as much fluid through grooming as they do through urinating.

Most cats have their own routine for grooming. Some spend up to one-third of their waking time pursuing a head-to-toe clean-up; others hardly bother. Most groom symmetrically and systematically, using their forepaws to clean the face and behind the ears, covering each foreleg with saliva before wiping the 'dirty' area several times in a circular motion from back to front. One scientist believes that cats secrete a special 'cleaning fluid', a detergent which cleans the fur. They certainly smell clean and don't seem to give off that dank smell associated with the less fastidious dog.

Contorting that extra-supple body, cats can reach almost all parts by twisting and leaning. But by far the best way to get to those inaccessible places around the ears, the back of the neck and under the chin is to get a friend to help. Friendly cats living together use mutual grooming as an integral part of their social bonding and will approach each other frequently for a wash-over, returning the favour later on. It is a way of creating a group smell which also includes humans and other animals such as dogs in the family which are also rubbed and washed. After we have patted and stroked our cats, they groom and 'taste' our personal scent and incorporate it into the communal scent picture discussed earlier.

Most short-haired cats are exceptionally proficient self-groomers and need little help from us to keep their coats in top condition. However, this is not to say we cannot join in and use grooming and touch, like all cats do, as part of the social bonding. Grooming using a brush or glove should be started as early as possible after weaning or getting a new kitten, thereby continuing the mother's role.

Ensuring that your attempts at grooming are accepted and enjoyed is especially vital in the case of some long-haired cats. Modern Persians are unlikely to survive in the wild because they are unable to keep their coats unmatted and an effective barrier against wet and cold. Also, many long-haired cats are bred with very flattened noses (e.g. the Peke-faced Persian) and probably find it difficult to breathe and groom at the same time, let alone hunt successfully. Many have been bred to have a more docile temperament and so are more tolerant of grooming. If you start young, the cat will be much more likely to accept your role, though whether it ever gets to relish regular brushing is another matter!

The worst thing to do is to delay tackling tangles, because it can then become a battle requiring the cat to be pinned down in a vice-like grip and attacked with brush, comb and scissors. The even more drastic scenario of a visit to the vet's for a general anaesthetic and a complete short-back-and-sides (down to the skin) may be necessary if long-hairs have not become accustomed to the essential grooming early on. The rule is: start young, and most cats will enjoy it. If you do have a long-haired cat, groom it at least once a day whether it looks as if it needs it or not.

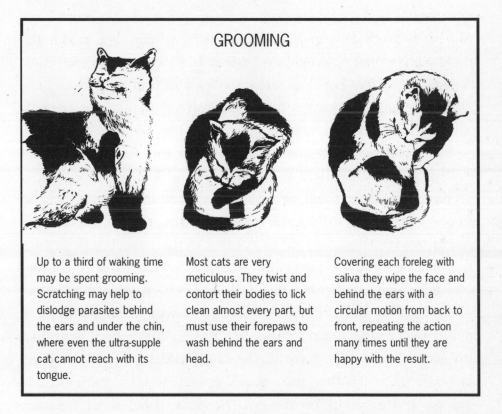

GROOMING

Up to a third of waking time may be spent grooming. Scratching may help to dislodge parasites behind the ears and under the chin, where even the ultra-supple cat cannot reach with its tongue.

Most cats are very meticulous. They twist and contort their bodies to lick clean almost every part, but must use their forepaws to wash behind the ears and head.

Covering each foreleg with saliva they wipe the face and behind the ears with a circular motion from back to front, repeating the action many times until they are happy with the result.

Cats are, of course, very individual in their likes and dislikes and some enjoy being brushed. Most enjoy being stroked and tickled around the head and some really revel in it. Almost all cats dislike their hair being brushed the wrong way, the reason being that it is likely to cause discomfort as their hair grows from the follicles at an angle specifically designed to allow the hair to lie flat on the skin and provide effective protection and warmth. Nerve cells in the skin tell the cat if the hairs are not lying flat – hence it doesn't enjoy being continually brushed the 'wrong way'. Of course, some cats enjoy being ruffled and tickled all over, and they may roll on to their back to expose their stomach, that vulnerable area which is usually otherwise well protected. Most accept a little tickling before grabbing your hand with their claws and teeth. They appear to be totally relaxed and enjoying themselves one minute, then snap out of the 'trance' and attack the hand quite ferociously. Americans have termed this the 'petting-and-biting syndrome' and it is in fact a basic defensive reflex: the cat first relaxes but then seems suddenly to find itself just a little too vulnerable and it snaps into a defensive display of aggression – often surprising itself as much as you.

But grooming is not just about hygiene and social bonding. It also has another function: to relieve tension. You may notice your cat having a quick groom after a fight or when it is upset. Cats also groom if faced with a dilemma, such as wanting to get out but being unable to open the cat-flap. The conflict in their minds is dealt with by what has been termed 'displacement activity' and is perhaps similar to when we scratch our heads or twiddle with our hair in moments of crisis. Excessive grooming can even occur if the cat is feeling under some form of constant stress – perhaps because it does not get on with another cat in the household, or if it is frustrated and bored. The cat then grooms and grooms to relax itself but can end up with a patchy broken coat or areas of baldness. Typically these patches occur down the centre of the back and at the top of the legs. Some cats even continue licking down to the skin and cause sore red patches to appear. These bizarre responses are discussed in more detail in chapter 7.

HUNTING

Hunting is essentially a simple response to a moving target, a response that is manifested because of a strongly motivated inborn urge, which is refined through acquired skills. These skills spring from early experiences when a mother cat will bring home half-dead victims for her kittens to play with, and they learn through trial and error the best methods by which to handle and kill the prey. Experimentation and reacting with its mother and litter-mates and its surroundings all improve and co-ordinate the cat's hunting abilities.

When hunting for potential prey the cat may make a search of fields, gardens or hedgerows with all senses on 'alert', and with frequent stops to investigate the immediate surroundings more thoroughly. This is referred to by biologists as a mobile, or M-hunting, strategy. Alternatively, hunting may take place in a known area, such as a 'run' used by mice in a hedgerow or the spot beneath an active birds' nest, where the cat waits patiently until something scurries past or attempts its first flight. This is known as the 'sit-and-wait', or S-hunting, strategy. In the latter case, the cat will approach, say, an entrance to a mouse burrow very cautiously so as not to announce its arrival and then sit motionless waiting for the first sign of activity. If you think about it, your cat is, in fact, adopting the S-hunting method when it sits by its food bowl waiting for you to fill it.

Many owners are rather disgusted at the way their pets appear to play with their prey, apparently torturing it to death rather than ending its days quickly. When prolonged, such play can indicate that when the cat was young it did not learn how to kill properly – perhaps because its mother didn't bring home half-dead mice to practise on. Hence the cat bats and tosses its victim only in the way it used to bat and toss its toys and play with its litter-mates and its mother's tail. The cat may also use the half-dead prey it has captured to practise its hunting abilities for the next time, safe in the knowledge that this particular victim can be killed and consumed at any time. Additionally, the well-fed pet has never had to learn to kill its prey quickly in order to relieve urgent great hunger. This begs the common question 'why do pet cats need to kill at all?' The answer is that not necessarily everything a cat kills is required for food. It may be used for the purposes of practice and simply as a toy or item of interest while it remains alive, moving and squeaking. Perhaps, as with many human hunters, and as seen in many dogs, there is a joy and an emotional high to be had in stalling and grappling with a victim, which feeling is lost once the prey is dead. The only way for a cat to maintain interest at this point is to make the victim 'come alive' again by throwing it around. Most cats quickly lose interest in dead prey. They either eat or abandon it and go off in pursuit of new live quarry. Prey hunting and killing occur independently of hunger in the cat.

Some owners try to give the birds and mice a warning by putting a bell on the cat's collar but often the cat learns how to move without making it tinkle. The cat may use the 'sit, wait-and-then-pounce' strategy outlined above and so the bell will not ring anyway until it is too late. One Garfield cartoon summed up the 'stop-him-hunting-with-a-bell' strategy by depicting the unfortunate cat tied to a bell the size of Big Ben!

Cats can be discouraged from bringing prey inside by the use of some of the aversion techniques entitled 'Acts of God' mentioned later on. However, these are unlikely to stop it hunting; it will simply take its catch elsewhere.

FEEDING

Left to fend for themselves, most cats will survive by employing a mixture of three methods: hunting, scavenging and accepting free hand-outs from local kind-hearted people. Depending on how closely the cat has previously lived with people, it may simply take what it can while keeping a safe distance or work its way into the hearts and hearth of a new family and there make the most of all they have to offer.

We can use food to bond with our cats, in this way fitting into the maternal role of food provider. Most pet cats never need to lift a paw to catch their supper; they need only gently to nudge their owners when they get peckish. So why do cats still hunt, even when there is food freely available at home?

The sensations of hunger and the urge to hunt are controlled by two different parts of the cat's brain, and although a full stomach may remove the urgency of going out to find a meal and make a warm snooze a preferable option, it does not override the desire to hunt. If a full stomach did stop the hunting instinct, we would probably see much less of the play behaviour we so love in our cats. Chasing after string or ping-pong balls or leaping at hanging teasers are all actions motivated by movement and the hunting instinct. Loss of our source of amusement and enjoyment from our playing cats would be sad indeed for most of us, although some owners of very proficient hunters may gladly sacrifice it to forgo the morning presentations of birds, mice, voles, frogs and a host of other victims their pet brings in.

Even old cats will revert to kittenish behaviour when faced with a ball of wool. The play sequences we find so entertaining in our kittens are merely practice for the real job of hunting, learning balance, speed and agility, even though the kittens never need to hunt to feed themselves.

Cats sometimes don't even eat what they catch, which seems to us even more of a waste, but this is characteristic of other carnivores too. Shrews and moles are usually rejected – they obviously do not taste good to cats. Some cats leave the gall bladder and some the small intestine of more palatable prey such as mice or birds, while others may eat the lot, fur, feathers and all. Bullet, our-black-and-white moggie, has the delightful habit of bringing in his catch, putting it on top of the food in his dish (dry food is always available for him), and eating it all, leaving the little gall bladder resting forlornly on top of the food. Just why cats bring in their prey is not known for sure. They may want to eat it in the safety of their den where they can relax and enjoy it free from the threat of theft from nearby rivals, or they may want to bring it in for us or other cats as a mother would for her kittens. Others seem to bring things in as 'gifts' for owners – much to their owner's non-delight, especially if the mice are not too badly injured by being caught to be able to scurry away to take up residence in the house. In one village called Felmersham, a study of what 70 cats brought home over the period of one year amounted to over 1,000 prey items. Some cats regularly brought in large numbers of prey, others hardly any at all.

So, what do our cats like to eat, how often do they want it, and how can we use food to build greater bonds with them? Choice of cat food is often based on human values and we are bombarded with advertising about meaty preparations and new flavours. These are usually geared to what we humans see as appetising. Of course, cans of mouse, sparrow or rat wouldn't be at all acceptable from our point of view, even though they may make the perfect meal for cats.

So can cats actually tell the difference between all the available varieties of food? As we saw in chapter 1, cats have a very strong sense of taste and smell, are sensitive to constituents of protein and can 'smell' fats. They are also very sensitive to the taste of water but not to the taste of 'sweet'. Therefore they probably know if they like a food just by smelling, and have no need actually to taste it. They decide if food is palatable using touch, smell and taste. It's common for them to like 'game' flavours. Our own cats are wild about rabbit and pheasant (and bacon and smoked salmon – they should be so lucky!) but are not particularly attracted by many other meats.

Given a choice of two equally palatable foods a cat is likely to choose the new flavour, which would suggest that cats like a variety of tastes and frequent change. However, if the cat is feeling stressed it will probably choose the more familiar taste, understandably clinging to the security and reassurance of a known quantity. Of course, each cat has its own preferences and whether those are for canned, dry or fresh foods it is best to accustom your cat to a variety of flavours and textures from an early age. To ensure that it not only gets a balanced diet, but is prevented from becoming addicted to any one particular flavour or type of food. The newer all-in-one dry diets which were initially introduced from the USA are made of a balanced combination of proteins and fats, minerals and vitamins from different, but consistent, sources, so variety is not so essential.

Appetite can be affected by how the food is prepared and by the cat's emotional state. Strange noises, new people or even a change of feeding bowl may cause loss of appetite. With their acute sense of taste and smell cats can very easily find stale food off-putting too. They like food best at 25 to 40°C – the body temperature at which their prey would normally be eaten – so don't feed cats with food directly from the fridge as its taste and smell diminish at low temperatures. Placing the food bowl too near the litter tray can also put cats off eating from one or using the other. We would probably react in the same way if our dinner was served up next to the lavatory. A position of safety, a little away from the madding crowd, is usually preferred as the cat can then take a relaxed bite out of the way of thieving dogs or curious children.

It is interesting to note that cats, with their origins in the desert, probably never ate fish as a result of their own hunting (there is only one specialist fish-catching cat species, which lives in India) and, apart from those occasions when they've snatched the odd goldfish from a tank or pond, cats have only ever had fish given to them by humans. The linking of cats and fish seems to have come about after the Second

World War, when the fact that fish was a ready and relatively cheap source of protein was used by pet-food manufacturers in their advertising. It was clearly a very successful campaign because the association has remained ever since. That is not to say that cats do not like fish, as many owners can testify.

Although cats do, of course, have their individual likes and dislikes when it comes to food, there is one thing that all cats have in common: they must have meat and cannot be fed on a vegetarian diet. While we and our dogs, being omnivorous, can decide, if we wish, to leave meat out of our diet without doing ourselves any harm (yes, dogs can live happily as vegetarians), a cat cannot survive on a diet without certain constituents, such as taurine, which are only available in meat. Diets advertised as vegetarian may have had taurine and othe nutrients vital to the cat added to them. If the diet has not got these additives it is inadequate and unsuitable for cats anyway. Cats do nibble grass and house plants, although probably as purgatives, vomiting them soon afterwards. Some cats even enjoy vegetables, but vegetables alone are not sufficient to fulfil all the cat's nutritional requirements.

Should cats be fed one large meal a day, several small ones, or should they have food available the whole time? Most dogs gulp down their food because in a pack of wild dogs or wolves there are many mouths to feed and a great deal of competition. In order to keep some semblance of order and avoid injury through fighting over every meal, they have a pack order which defines when each is allowed to charge in and feed. First the top dog eats until he has had enough and is then followed by the others, in rank order, until each has had its fill. The bottom dog often has to hang around and hope there will be some left. Cats, being solitary hunters, are generally less voracious and, if food is freely available, will nibble frequently rather than devour at one go. Results of feeding studies show that cats given free choice will eat up to twenty times during a 24-hour period. This appears to mimic natural feeding patterns. Mice, being the domestic cat's most common prey, provide about 30 kilocalories of energy each. An active cat weighing 3.5kg expends approximately 300 kilocalories of energy a day – about ten mice or ten meals a day. However, our cats do not have to catch mice to survive and the needs of most healthy, mature house cats that are not pregnant or lactating can be met, if the cat food is of good quality, by feeding them one meal a day, though they still hunt despite this.

But feeding is about more than just getting enough calories and nutrients: feeding is a very important bond builder between cat and owner. So, whichever type of food you select, mealtimes can be used as a period of interaction and communication. When the cat asks for food we can respond, and receive a very warm response in return.

When we call at mealtimes, the cat will come running. Many will come running *before* we call because they have picked up cues from our behaviour that feeding time is approaching. Finding the tin-opener, washing its bowl or the click of a certain cupboard door are all clues and the cat usually appears eagerly, out of the blue. A conversation of miaows and replies goes on throughout the bowl-filling procedure and the cat will rub and mark its owner and even the food bowl to give encouragement and prevent distractions which might delay the process. After a quick sniff to check out the menu, the cat then tucks in to the food, its soliciting, appealing behaviour now forgotten and the owner no longer required!

While a cat can be fed enough in one or two meals or provided with dry food to eat *ad lib*, behaviourists advise that, especially in the first year you own the cat, you feed it little and often. Interacting over each small meal enables you to bond very closely with your cat and continue the original relationship that it had with its mother. At feeding times both you and your cat make an effort to interact and so you can easily take over and reinforce that maternal role in the cat's life and maintain your friendship. This is the essence of the owner's successful relationship with the cat and it is one of the main reasons why they can relax and enjoy our companionship. Once these bonds have been established cats can be fed *ad lib* or larger, less frequent meals.

Unlike their more greedy canine cousins, cats usually eat only what they need and are much less prone to obesity. There are considerable variations in the size of cats, even within breeds, so it is not easy to determine an optimum weight. If you think your cat is overweight, one indication to look for is an 'apron' of skin and fat which hangs just beneath its belly – it may swing as it walks along! Often cats, especially the Orientals, may not put on weight all over, but instead carry the 'extra' slung below. Your vet will be able to advise you if a diet is necessary and how to go about weight reduction in a safe and practical way. Many cats seem able to eat heaps of food and not become an ounce overweight.

A CLEAN JOB!

The cat is inherently clean in its habits. Not only does it spend up to a third of its waking time grooming, but its toilet habits are usually immaculate as well as modest, and this is another reason that cats are such successful pets. The cat will usually choose a spot that is protected on as many sides as possible, such as a quiet corner of the garden or under a hedge, where it can safely adopt that most vulnerable of squatting positions. It will dig the soil, having chosen a patch where it is fairly soft (new seed-beds are a favourite), defecate or urinate in the hole and, after sniffing the area, fill in the hole with soil. Conveniently for us, it's all done outside and not usually in our own garden, and it's not even something we need to think about. However, a cat-litter industry worth over £40 million per year indicates that many people do have to consider what happens at 'the other end' and they put a lot of effort into carrying bags of litter home for the cat to have somewhere to relieve itself.

Cats may use a litter tray indoors for only a short while, such as when they are too young and unvaccinated, and would be vulnerable to infection outdoors; when they are too ill, or have just become too old and frail to go out on patrol. Many cats are now kept permanently indoors in cities because of the outdoor dangers of traffic. Others, like Flirty and Bean and many such single-coated breeds, don't like putting their noses outside in any type of cold weather, let alone trying to dig in frozen ground. They prefer to use a litter tray indoors most of the time. Bullet, like most moggies, goes out come hell or high water and wouldn't dream of using an indoor tray.

When we take kittens on at between six and twelve weeks of age (or when they are over twelve weeks old in the case of pedigrees), they usually have no problems using a litter tray and we pride ourselves on our success in litter-training them. In fact all the work has already been done for us by their mother and, provided the kitten knows where the tray is and can get into it, everything usually goes smoothly for the new kitten owner. Tiny kittens learn from birth never to soil their bed. To begin with they are unable to urinate or defecate without stimulation of the abdomen and genital area by their mother's tongue. This reflex action of the kittens can remain operational until they are about five weeks old, though most can urinate and defecate voluntarily by three weeks. As the kittens start to wander and explore outside the nest their mother continues to look after their toilet needs. She carries them away from the nest, training them for life that the nest must not be soiled. Good mothers teach their kittens well, and each generation teaches the next.

As the kittens play they paw at the ground, explore the litter and learn about things around them. They instinctively rake and scratch in soft loose material. They also learn a great deal from watching and imitating their mother, and soon associate the smell of the litter tray with toilet activities. By the time they are weaned at about six weeks old, the kittens have formed the habit of using the litter themselves. This association usually transfers well to using soil outdoors at a later date when the kitten's vaccinations are completed.

It may seem that we have little part to play. Placing the kitten on the litter after it has fed or just woken up, when it is most likely to need to relieve itself, is usually enough to ensure that its enquiring little mind associates the two functions. But we should have a thought for just where we place the litter tray and what we put in it. Cats usually try to find a sheltered quiet spot to 'go to the toilet', so an open tray in a passageway or by the dog's bed may not be well received and, indeed be boycotted by a more nervous cat. Placed in a corner, a tray covered with a box (with a cat-sized hole cut in one end, of course) or a ready-made covered tray may make the cat feel more secure and content to use it. A cover will also prevent the mess created by an 'over-enthusiastic digger' intent on moving most of the litter out and over the side of the tray.

Cats like to be able to dig and so newspaper in the tray, as is sometimes used to house-train puppies, is not advisable. Also, the type of litter offered is more important to some cats than to others. Being originally semi-desert creatures, cats will enthusiastically use sand or fine grain material if it is available – as most builders will know on their return to cement-making after a weekend to find added bonuses in their sandpile! Many parents too find the kids' sandpit is frequently used by local cats. You can use this preference to your own advantage if you have recently dug the garden as the sand will be highly attractive to the entire cat population of the street. A pile of sand in the corner of the garden will probably be even more irresistible than newly dug soil and you may save the seed-beds from the ravages of local cats. I say probably because, typically, there is always one cat that breaks the rules. In his book *Travels with Tchaikovsky, a Tale of a Cat*, about his world-wide travels with his cat Tchaikovsky, Ray Hudson recounts how Tchaikovsky had to be given a litter tray in the middle of the desert because he wouldn't use the sand. So much for the influence of Egyptian ancestors!

Litter for kittens must be disease-free and this cannot be guaranteed with sand unless a sterilised variety is used, sometimes available from aquatic shops. The most commonly used litters are made from Fuller's earth, a granulated dry clay substance which forms clumps when wet. Other litters include the pelleted wood-chip types, which are much lighter to carry, and some finer-grain clay granules. Cats which live permanently indoors may have much softer pads than their outdoor relatives who have hardened up their pads on concrete walkways and through long nights spent on the tiles. Indoor cats may find the larger pelleted types of litter a bit uncomfortable to stand on and prefer the fine-grain varieties.

The type of litter determines how often you need to attend to it. Some require frequent replacement; others 'clump' allowing solids and even liquid waste to be lifted out with a scoop, which leaves the remaining litter available for further use. Removal of waste needs to be done regularly, before the 'clump' is raked around by the cat at the next use. Most litter needs picking over at least once a day in order to prevent excessive soiling (which puts the cat off using the tray) while allowing it to retain the smell of a latrine and so keeping up the cat's toilet association with it. This is particularly important when bonding a new kitten to the litter.

Cats are usually very fastidious and 'accidents' are rare. Giving this mundane but necessary practicality some thought from the cat's point of view can help the cat feel more relaxed and prevent problems occurring. If they do occur it is usually because of illness or some upset and tackling the problem is covered in more detail in chapter 7.

TIPS: GROWING CLOSER TO YOUR CAT

- Greet your cat face to face and exchange scents by rubbing and stroking.

- Keep hands out of the way with a nervous cat until it has relaxed.

- A single-syllable name is learned more easily by cats. Say it in the same tone each time so it knows it's being talked to.

- Touch is important so give lots of attention.

- Begin grooming early with young kittens and never let it turn into a battle.

- Don't feed straight from the fridge. Let the food warm up to room temperature before serving.

- Provide a warm safe bed.

- Feed frequently when getting to know your cat in order to build up bonds.

- Remove uneaten food. The smell of rancid fat is very off-putting.

- Cats cannot be vegetarian.

- Position the litter tray in a safe place and find a litter your cat likes.

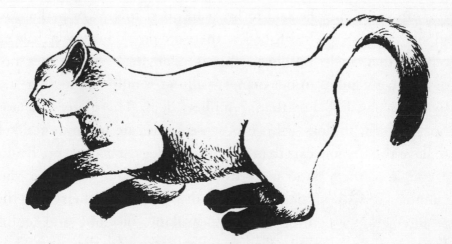

4

Exploring the Relationship

So far, we have looked at how we can use our knowledge of the cat's behaviour and preferences to communicate with it and to pander to what it wants. But we too want something from the relationship – what do we want, and how can we ensure we too are satisfied with the feedback?

Cats could now be said to be 'man's best friend', or perhaps we should say 'woman's best friend' as there are probably more female cat owners than male (although this is changing too). Felines now outnumber canines in our homes – almost 8 million cats share our lives compared to less than 7 million dogs. There are still more households with dogs – this is because homes are much more likely to have two or more cats than two or more dogs. If cats get on, having a couple of them is no more trouble than having one. The same cannot be said of dogs in general because of their size, the requirement for control when out walking, or noise and fouling factors – two dogs are usually a great deal more work than one. Those non-cat-loving folk who have never shared their lives with a feline may well ask how this independent, aloof creature that doesn't even do as it is told has found a place in so many hearts and homes. Indeed, we ailurophiles (cat lovers) do seem to be an all-or-nothing type of people. Few are merely lukewarm about cats – you either love 'em or you hate 'em! Over 50 per cent of cat owners let their pet sleep on their bed and many even buy it birthday and Christmas presents, though quite what such good fortune means to a cat is unknown. A few people fear cats and, to these ailurophobes, being in a room with a cat can bring on the same sort of feelings of terror that others would experience in the presence of spiders or snakes, and, like most phobias, for the same illogical or inexplicable reasons. What is that indefinable quality we love or hate so much? How have cats wheedled their way so closely into our lives?

THE ADAPTABLE CAT

Originally, cats probably became associated with man in ancient Egypt, where they proved to be useful as vermin controllers around grain and food stores, but rodent-catching is the last thing most of us now expect or want from our pet cats. In fact we don't, on first sight, really 'expect' anything except their companionship. Cats have never been manipulated for our needs in the way dogs have, to work alongside us herding the sheep, flushing game or carrying out guard duties, for example. We may, through selective breeding, now have many pedigree strains of various colours and coat types, but the vast majority of cat owners still keep a 'good old moggie'.

So, how has the cat managed, without changing, to outdo the dog in the popularity stakes? Both the cat and the dog are fundamentally as they have been for thousands of years. The dog is still 85 per cent wolf, despite what we have done to a few of its genes, and the cat, according to some of the latest theories, is actually genetically no different from the African Wild Cat; it has merely adapted itself to live with us anywhere in the world we have taken it. What *has* changed dramatically is our society, and the way we live and work. The cat or dog has had to fit in with all our changes. It is a credit to the flexibility of both species that they have fitted so well thus far, but our systems are becoming more and more urbanised and difficult for them to cope with. Many of our own roles, especially those of women, have widened in terms of choice of careers, and been dictated by the need to work in order to pay a large mortgage, but this has meant there is less scope for the role of traditional mum at home with kids, cat and dog. Obviously many women and men do stay at home with the family, but it is frequently only for a few years rather than for the pet's lifetime. Often young children are taken to a minder and collected by a parent on the way home from work, and the animals must spend the day alone.

In the past century man has moved in from the country to live where work was concentrated in urban and suburban centres. Over 80 per cent of us now live in these areas, albeit suburbia in the UK is still relatively green with parks and heathland. This is not to say people do not crave for a more rural life; it's just that the opportunity for employment lies in more populated areas, where there are also, of course, more facilities for leisure, more schools and public transport.

Although we are slowly moving towards flexi-time and working from home (technology is being developed at an amazing rate), it will be many years yet before most people will be let off the daily journey to and from work. Both dogs and cats can give us a link back to the nature we hanker after, as well as a chance to interact and unwind.

While a farm or country house with a large garden and easy access to fields is ideal for keeping dogs, many also live happily in suburban or city areas with regular supervised access to exercise areas. However, the more dense the population and the fewer the areas available for walks, the more effort owners have to put into finding the dog a place to exercise and, more basically, to defecate or urinate. While cats can let themselves in and out through a cat-flap, or use a litter tray, dogs still require us to let them out and so the time for which they can be left unattended is limited.

Dogs are basically pack animals, enjoying and needing the company of other dogs or, as equally acceptable substitutes, humans. Many find it difficult to cope on their own and when their anxiety overspills it leads to chewing, barking or howling, which causes problems for their owners and the neighbours. Cats are more solitary creatures, although they do socialise with, and enjoy the company of, other cats. Separation anxiety rarely occurs, although the occasional 'dog-like' Oriental cat does seem to suffer if it is closely attached to an owner who leaves it at home alone.

So, in an environment where no one is at home all day, where the working day starts early and ends late, the cat is a more convenient option as a pet than the dog.

It should also be said that the present climate is not particularly pro-dog and owners of large or fierce-looking dogs are probably aware of the element of fear shown by other people towards their pet when they walk in local parks. With all the publicity about bites and injuries caused by dogs, owners must be more careful than ever – even if they have a friendly dog. Those who keep dogs with a dubious temperament, especially in areas of high population density, have to be particularly wary. Many now muzzle their pets, and some are required to by law or face the threat of litigation should an incident occur. Since we are legally liable for the actions of our dogs, as well as insuring them against veterinary fees many owners have to take out third-party liability against injury or accident caused by their dogs. There is also the health outcry over dogs messing in parks and on pavements, a particular problem where children play. Cats may use the neighbour's garden as a toilet, but they usually get away with it because they cover their deposits afterwards.

On the other hand although, legally speaking, cats cannot be considered to trespass or be held responsible for causing an accident, negligent cat owners can be held liable for injury against a person, for example, if they knew the cat was aggressive but let it out anyway – in practice, something very difficult to prove. Cats offer us a low-risk relationship: we need make few demands on their behaviour but they give us high emotional rewards.

Dogs may be suited to a wide range of types of home, but cats are more adaptable and can live in the wild near to man in feral colonies with minimal or no human contact, or, at the other extreme, lead an entirely indoor life in a tenth-floor flat. So, in modern-day life, with all its pressures, rules and limited space, the cat has proved it can be 'a pet for all seasons' and suit a range of lifestyles – from city flat to country manor.

WHAT DO WE NEED FROM OUR CATS?

In surveys which ask people why they keep cats, the respondents have usually replied along these lines: 'companionship, cleanliness, love . . . to enjoy the perfection of the feline form and personality . . . and because the cat is so convenient and easy to keep'. Dog owners will also cite companionship, mentioning the added benefit of home- and self-protection. They may not admit to the status-raising or ego-boosting reasons behind, for example, owning a large fierce-looking Rottweiler or a beautiful, graceful and expensive-looking Afghan. The cat is a much more personal pet. A few may be kept as status symbols because of the luxury of their coats or the length of their lineage; but most, pedigree or moggie, long- or short-haired, are loved each as much as the other and, despite great variations in their individual characters, for mostly the same reasons. While the owner of a toy poodle may not understand why someone would want to keep a mastiff (or vice versa), cat owners understand the lure of the cat species as a whole, whatever the shape, size or hair length.

The innumerable combinations of cat and human personality make each relationship unique and although there are always tales of special cats which act more like dogs – those that travel all over the country with their owners in cars or trains, or those that have saved families from fire – most cats and their owners enjoy a very similar relationship. While there are many ways in which we see the cat and in which we enjoy its company, the intensity of the relationship often depends on how much we 'need' it in our lives.

FAMILY CATS

Many cats live with human families. They may have started off with a single owner, but as the life expectancy of a cat now averages twelve years, and many live more than twenty years, they are often joined later by their owner's spouse and then any number of children, dogs, ponies, rabbits, gerbils, goldfish and, in 35 per cent of houses, other cats too. While the children are young there is usually someone at home and the cat moves about as it wishes, finding high and safe niches when the kids are at the 'poke-and-grab' stage and the puppy has not yet learned that the cat is boss. The cat comes and goes as it likes and is fairly independent in its actions. Nobody actually has time to notice all its movements or to try to follow up every miaow, so the cat learns to muck in and make the best of things or keep out of the way.

The children also learn to interact with their pets, to look after them and be responsible for the routine operations involved in feeding and grooming. The cat may also be the child's 'only friend' in those times when he or she feels misunderstood or rejected and is contemplating a 'running away plan', though fitting the cat into the rucksack might prove a bit more difficult and may be just the reason to stay at home and live through the problems. Usually, confiding in the cat by cuddling and relaxing together is enough for the child to get his worries out of his system and soon everything gets back to normal. Children today have fewer chances to integrate even with other children and animals in their street because of the understandable worries about their safety outside. As part of a gang of children growing up in a small town in Ireland, I vividly remember the dogs that would wander past each day on their travels around town – they would drop in for a biscuit and play with the kids and then go on their way. Cats too moved among the houses – having a cat give birth to kittens in the coal bunker was exciting but by no means extraordinary. Of course, there were many more stray dogs and cats then and many kittens died from accident and disease, observation of which, though painful, did teach us kids about death. In those days there was also more space in which to allow animals such freedoms. This is a completely different scenario from that of the crowded towns of today – at least, it is in England, but in many places in Ireland it probably hasn't changed much, despite tighter controls on stray dogs.

A Cambridge study which compared groups of children with and without pets found that those with pets in the family had better relationships with their parents, and a happier home life than those with no pets. According to the survey, the more pets, the better the family relationship. This is perhaps because pets help to spread the responsibilities of entertainment, or they decrease the competition for parental involvement in everything, spreading the need for attention and love over a larger number of individuals, be they people or animals. Pets also provide a common outlet for interaction and reduce sibling rivalries.

CITY CATS

In a village the feeling of 'home' can sometimes be brought on by turning the corner into your street or area because the people there are part of a community known to you. However, in modern cities many people do not even know who lives in the house next door or the flat above or below, and 'home' is concentrated into one small house or apartment, or one bed-sit room and its contents. Most childless couples living in and around cities are also likely to be out at work all day and the first person to return must open up the quiet and unwelcoming house and begin to make it feel like home in order to relax. After a frustrating train or tube journey amid thousands of other people also desperate to get home, or a fight through the rush-hour traffic, the stresses and strains of the workday need to be released.

This is where the cat comes into its own and this is one of the times when we can use the cat to help us. Inside that small bundle of fur hides a perfect therapist. Even curled up on the sofa amongst the cushions it glows with contentment and complete relaxation. A welcome mew is all that's necessary to break the ice and bring you straight into 'home and family feeling'. Where the flat is owned by a single person – and the number of single-person households is rising sharply each year – the cat is even more treasured and valued for its companionship, intelligence and ease of management. It also provides a taster of a return to nature, keeping us in touch with the basic values of life and simple pleasures. The cat fulfils a need to care and to interact without question, to express emotion and receive love in return. It is among such owners that the number of cats is rising rapidly and where the 'need' for the cat is strong. Such cats often have to be very adaptable because owners may expect to interact and be given lots of attention when home in the morning and evening, but then to leave the cat on its own for the rest of the day.

In the countryside the average cat density is one cat per twenty to twenty-five acres. In London and other cities this may rise to one cat per 0.02 acres – an amazing 1,000 times increase in density. That 'solitary' hunter is likely to bump into an awful lot of others on his travels across gardens, and a new cat moving into the area may have a tough time creating a niche for himself because of the lack of space and the number of other cats to which he must make himself known. Along with increased cat density comes the added danger of more people, dogs and traffic and this is the reason why many people in the city keep their precious pets permanently inside the home.

PERMANENTLY INDOOR CATS

Permanently indoor cats certainly are not able to exercise that most feline of prerogatives – to leave and move on if they don't like the present conditions. Thus they are under much more pressure to conform to human expectations than their easy-come, easy-go relatives. Because they live in close proximity with their owners, all aspects of their behaviour can be observed and their habits and preferences carefully noted. The cat must learn to order its day and week around its owners – after all, things only happen when the owners are home. The owners should be aware of the huge impact they have on their cat's life when they are together and the responsibilities that go with such control. Dr Dennis Turner, who has carried out in Switzerland much of the research on cat behaviour and the cat/person relationship, found that owners of exclusively indoor cats were less willing to accept cats as being independent creatures. They tended to reject the idea of a cat deciding for itself the nature of its lifestyle and relations with its human family or fellow house cats. Nevertheless, owners of indoor cats were more tolerant of their cats' less acceptable behaviours, such as furniture scratching.

However, animal behaviour therapists see a disproportionate number of indoor cats compared with 'free-range' pets. That they do so is because the confined cat must necessarily act out its entire behavioural repertoire indoors, whereas the free-range cat would normally spend a portion of its day outside playing 'outdoor hunter/territory patroller'. The latter role needs vigour and great concentration and allows the cat to 'work off' both mental and physical energy and to revert to instinctive behaviour. The type of breed may be an important factor in choosing a cat that will be confined to the home. The more energetic breeds which bond closely to their owners, such as the Siamese and Burmese, may not be the best choice for a solitary indoor cat in a small flat. A quieter, more 'laid-back' breed or a good old moggie may fit the bill better and cope better on its own, conforming to the adage 'what you haven't had, you won't miss'.

For the same reasons indoor cats need lots of entertainment and novel objects to investigate to prevent them from becoming bored and to exercise their highly developed predatory and exploratory natures. Getting two kittens instead of one means that they can amuse and keep each other company and share the responsibility for entertaining their owner too. This is not, of course, the sole answer to keeping the indoor cat fulfilled or content, but the advantages it brings are certainly worth considering.

Put yourself in the cat's place. Imagine that your entire world is a two-roomed flat. You will notice any slight change in the positioning of a picture or the placing of a chair – the cat knows the profile of the flat intimately, both physically and by smell. To a sensitive cat, changes, like a new carpet, or even smells brought in on its owner's shoes, can be very upsetting. Keeping the cat interested and able to cope with changes by bringing in a constant supply of new toys and visitors may help ensure that it does not become oversensitive or over-reactive to small, unimportant alterations in its environment.

Any problems or changes in the cat's behaviour are immediately noted and the owner often takes them on as 'his fault'; he assumes they are caused by something he did, or something he didn't provide for the cat. Indeed some problems can be very dramatic, such as when toilet habits break down in a one-room flat. The cat may also become very close – or even over-bonded to its owner and suffer during absences.

Cats live for the moment and initially have no perception of the working week and playing weekend, although they will alter their daily routine when the owner is around. An owner may complicate matters even further for an indoor cat by taking it to a country home for the weekend and allowing it to go outside. (Two days of good fun paid for by five days of frustration!) Although the cat may not have been outside before, this rarely poses a problem and even cats that are not introduced to the big wide world until they are seven or eight years old adapt amazingly well, taking to it like ducks to water.

However, the trouble begins when the cat goes back to enclosure in the flat for the week and is expected to return to its former ways, away from the pleasures and excitements of the outdoor world. One moment it is kept in, the next allowed out, where it reverts to a wild instinctive personality. Some cats cope well; others may not. One such weekend-country cat began to over-groom; to lick his paws and pull his hair out until it began to remove his skin too – a condition known as self-mutilation, and caused by 'stress', with all its intricate meanings and causes. The cat's owners no doubt thought they were giving him a huge treat by taking him with them to the country but they were unable to understand the frustrations and uncertainties it made in his life at other times.

Usually though, if kept amused with toys and new games and given lots of play to tax its energy (a feline companion will provide much of this), a cat will cope quite happily with being kept indoors – surely a quite extraordinary adaptation for that tiger in our living room.

WOMEN AND CATS – CHILD SUBSTITUTES?

Women are often the most ardent cat lovers. Many men also own cats, but may not have quite the same 'over the top' (as some may call it!) enthusiasm for their pets. This has been explained to some extent by describing men as having much more of a group mentality or group hunting instinct (like dogs) and their not identifying with a solitary hunter like the cat, but this simplicity is probably to insult many cat-loving men. Women tend to devote more time to affection and attention and to nurturing the cat than men, and so perhaps they receive more response from their cats than men do. Often it is the woman who does most of the cleaning and feeding and so she gets to know all aspects of the cat's life better.

The notion of the cat as a child substitute is not new and it is generally raised on account of the number of women who keep cats, and their reaction to them. The relative size of the cat's eyes has already been mentioned as being part of its babylike appeal. This, together with the cat's enjoyment of a close warm cuddle, of course encourages the comparison. Cats are often likened to babies, but usually by younger or older generations, not by the average working 20- to 55-year-old woman. No doubt some cats are child substitutes, though this is more often the case in older, wealthier women who may be lonely and lavish much love and attention on their pet or pets – but why not? The providing of love gives a sense of self-esteem and for most people this is what the cat fulfils, a need to be caring and a need to be needed.

SPECIAL THERAPY

In some cases, being needed is what actually keeps people going. Many older people, unable to get out because of physical disabilities or fear of attack, literally have their cat as their only companion. As well as providing love and companionship, the cat gives them a purpose and meaning to a life that may have little else to offer. This feeling of love and worth can be shared with the cat by it merely choosing to sit and purr on that person's lap, coming when called and 'replying' to a conversation. Stroking is known to decrease stress and high blood pressure and produces a feeling of well-being. Indeed, it has been discovered at some forward-thinking institutions that the staff could help the residents to relax and to cutback on their sleeping pills by getting them to caress and stroke a cat.

Many institutions that care for the elderly or for the mentally ill have also seen startling improvements in people's self-esteem and desire to interact when a cat or dog has visited. Schemes have been set up for just this purpose and, in many cases, there is now a resident cat, which usually thrives in the warmth not only of the high ambient temperature offered for the elderly but in the love of the people there and the overwhelming number of laps awaiting its presence. There are many stories of people who have been institutionalised, who have retreated into themselves and are only beginning to speak for the first time after years of treatment, using the cat as an intermediary in conversations. Many of these cats too live up to their therapeutic role and seem to know who is in need of attention or a little love – that sixth sense again?

TRADITIONAL FARM CATS

There are still a few traditional farm cats about and their generations populate the outbuildings and barns and feed on rodents – those that are not controlled by poison or traps. Nowadays there is little chance of a free meal from the milk-maid who, folklore has it, as she hand-milked the cows would squirt some milk into a dip in the ground for the cat to lap up. Modern noisy milking parlours are not such attractive places for cats and spillage is rare. Immediate disinfection in the parlour and a rapid throughput of cattle means it is not even a safe place for a small cat to linger. Colonies of farm/feral cats do still congregate around farms but are often also controlled as vermin themselves, while a favoured few may be allowed in the farmhouse and fed as pets, just as they have been for centuries.

FERAL CATS

Feral means 'domestic gone wild'. Feral cats are simply pet cats, or their descendants, living away from the direct care of man and out of his home. They are not a different species like the true Scottish Wildcat. Often they are cats which have been dumped or become lost or are succeeding generations of cats living wild. Feral cats often form colonies, numbering from a few to hundreds, grouped around a source of food and shelter. Docklands, hotels, industrial and hospital grounds are common sites. They hunt, scavenge and generally find enough pickings to survive, though they are often helped by people who regularly feed them. These meal providers are usually devoted individuals who, once or twice each day, make their way to the colony with bags or trolleys of food and are greeted enthusiastically by 'their' cats. The relationship between feeder and cats can be an extremely friendly one where the cats allow themselves to be stroked and even picked up; in other cases some will not move forward until food has been put down and all humans have left the scene.

The cats certainly need the feeder, especially if they come to rely on the regular meals, but the needs and motivation of the feeder are more complex. Often feeders are old ladies, stretching out their small pension beyond their own requirements in order to buy food for what they see as 'their' cats. That they need to be depended on and enjoy exercising their strong sense of responsibility goes some way to explaining the reasons for their actions. The exclusivity of the cats' warm welcome and interactions and the feeders' subsequent chance to view the cats (which run from most other people) are also a great reward for such huge efforts.

HOW DOES THE CAT SEE US?

As we have seen, we keep cats for many reasons. However, how does the cat see us and does it have a range of requirements from us as we do from it?

When cats are with us they are usually in 'indoor' mode. Safe in their den, they are relaxed – they need not stir themselves to hunt for food if they don't feel like it; they can settle into accepting warmth, comfort, companionship and food, just as they did as kittens with their mothers. With us, they re-enact the time in their lives when they were most sociable – kittenhood. They demand attention and cry for food, and more often than not are rewarded for their efforts. They then respond by purring and are cuddled all the more, a rewarding, positive feedback for both pet and owner. We cradle cats in our laps like a mother cat would curl around her sucking kittens and their pleasure often spills over into kneading with their paws, just as they did to stimulate milk flow from their mother, and even into dribbling and sucking at skin or clothes. Playing with pieces of string animated to encourage a game, we mimic the role of the mother bringing home prey disabled just enough for kittens to play with and with which she educated them in the skills of hunting.

It is obviously too simplistic to describe the relationship as purely a kitten/nurturing mother one. Perhaps it would be more accurate to say that the relationship we enjoy gives the cat the confidence to relax and allows it to exhibit the behaviour we would rarely see otherwise and which it only revealed to its mother and occasionally to other cats it feels completely at ease with. There is no competition in the relationship with us – the cat does not see us as another cat but as security and a provider. Cats which get on with each other do play, even when they are quite old. However, often this play is chase or wrestling, and even in grooming (if you watch carefully) you will see that there is a level of control and it is important who grooms and who is groomed in terms of status. It is usually the stronger cat which does the grooming – the fact that the other cat allows this means he is giving in a little and allowing this intimate communication. With us there is seldom competition – they relax and let down their guard. Cats also respond to us as they would a mate – that characteristic 'bum-in-the-air' stance (the reason why one of our Siamese kittens was given the name Flirty Bottom!) is a sexual response of willingness. Grooming is enjoyed throughout the cat's life way past kittenhood and mutual grooming between cats which are friends and between people and their cats is obviously very enjoyable. However, the most appealing and rewarding gestures, from our point of view, are the ones in which the cat re-enacts those juvenile behaviours such as kneading and purring that only its mother would have been privy to.

TIPS: RELAXING WITH YOUR CAT

Now you know what makes your cat 'tick', how can it help you to relax, to cope and to make the relationship enjoyable for both parties?

- Respond to your cat's desire for contact and enjoy making him happy.

- Give him short but frequent bouts of attention to keep both of you fresh and keen for contact.

- Use your cat to relax after a hard day's work or vigorous activity. Stroking slows the heart and lowers blood pressure.

- Chat your worries away and confide your secrets – he'll never tell and is sensitive to your moods too.

- Enjoy a loving cuddle and a feeling of calm.

- Indulge him with the odd treat and revel in the feedback he gives.

- Accept and admire the wild side of your cat. It provides a taste of nature and watching his outdoor activities, even in built-up areas, gives an insight into his exciting world.

- Remember that permanently indoor cats need stimulation and play and lots of attention so they can express their whole behavioural repertoire within four walls. You'll enjoy the privilege of seeing and providing outlets for his whole personality.

- Invite an elderly neighbour round. Your cat makes you feel good so perhaps it will give the neighbour enjoyment too.

Cat Characters

Any cat owner will tell you that his or her cat is very much an individual, a 'real character', different from other cats, with its own likes and dislikes, habits and peculiarities. If all cats looked exactly alike, could you tell yours from the others just by its behaviour? This behaviour is an expression of its individual personality and style. Because we live so closely with our pets we can identify complex features of their behaviour. By 'domesticating' cats, by providing for them, we may have removed the pressure on them to be successful hunters, and allowed other behavioural traits to develop more fully. An impression of a wide range of very distinctive individuals emerges if you chat to cat owners, but what makes a cat become a specific personality? Of course, we're still trying to fathom that question in our own species and it forms the basis of the nature versus nurture argument. So do we know how much influence a cat's early environment has on it, which characteristics are genetically controlled through inheritance from its parents and ancestors and whether these can also be affected by the cat's experiences? Is a cat timid because its parents were, or because it had an 'unhappy kittenhood'? There are many factors which contribute to the cat's character – how bold it is, how good a hunter and, most importantly, how friendly it is to us, to cats and people.

IT'S IN THE GENES

Does the 'friendliness' of the kittens' parents affect how friendly the kittens become? The little work that has been done on this subject has led scientists to conclude that friendly fathers produce friendly kittens, even though in the research, the kittens had never met their fathers. Thus they had inherited the potential to be friendly, not learned to be so by observation. No doubt friendly mothers are also a great asset, as kittens would learn much by watching and copying her responses. In an experiment in which a cat had to press a lever to get food, kittens that saw their mother successfully complete the test were also successful. Those which didn't observe their mother were not able to work the problem out on their own. So if a mother cat is relaxed and confidently interacts with the people around her, the kittens are likely to copy her. Of course, many influencing factors overlap where mothers are concerned – a friendly mother, as well as passing on in her genes some tendency to be friendly, will probably be more open to allowing her kittens to be handled and to integrate successfully into her human family.

The health of the parents, especially the mother, has an effect not only on the physical growth of the kittens but on their psychological development. An undernourished mother is going to have a much harder time successfully rearing all her kittens and she will probably be less active and attentive in other motherly duties such as teaching social skills. Kittens of undernourished mothers learn less, more slowly. They can also show antisocial behaviour to other cats, abnormal fear and aggression, and poor development in physical skills – such as mastering the techniques of balance. Perhaps because the kittens have had to spend more time sucking and competing with brothers and sisters for limited milk supplies, they have not learned to relax or play. Early learning is vital to all the skills, both social and physical, that the cat will need in later life.

PERSONALITY TYPES

A researcher in the USA has recognised two different types of feline character within the groups of kittens she has studied. She categorised (no pun intended) them as being those which react quite vigorously to situations and are what we could call excitable or nervous, and those which are much more relaxed and quiet in their attitude to life and its challenges. Breeders may like to do their own studies on their cats' litters and see if they can identify these two characteristics among them.

Current work on the personality of the adult cat also points to the likelihood of there being two different and distinct types. Type one, the one most of us would probably prefer as a pet, needs lots of social contact with both people and other cats and is relaxed in their company. The second type seems only to enjoy the company of one or two members of its human family. It 'puts up' with social interaction with other people, and often doesn't form bonds with other cats either.

While many people enjoy the first type of character and think of this as a normal friendly cat, others are disappointed with their seeming inability to bond with type two. This is further compounded by the cat becoming more aloof the more the owner tries to nurture it and make it enjoy their company. Type two cats don't usually initiate contact and will distance themselves even further if more effort goes into trying to cement the relationship. Owners of such cats will probably recognise this profile and may be relieved to discover that their unloving cat is not that way because of anything they themselves did or did not do. Nevertheless, this behaviour in type two may explain why some people who have previously had a cat do not want to replace it and seem to disbelieve tales of close and loving cat/people relationships, summing cats up as aloof, independent and unrewarding.

Whether these two types of cats are the same two types that were identified in the American study on kittens is not yet known. What actually caused these adults to be friendly or not? Was it early experience or lack of handling in those vital early few weeks, or are they merely the over-reactive kittens grown up? Clearly there has to be a great deal more study and not just with kittens, but through watching them as they grow up and seeing what type of character they develop into. While it is easy for us to sit in our armchairs and say 'it's obvious', it is very difficult to collect information about normal cat behaviour when 'normal behaviour within a family' is what we are talking about. Asking owners to fill in questionnaires always results in a great deal of subjective reporting, but having an observer in the house often means that the cat does not act normally – especially if it is of the reactive type. So we may be a long way from solving all the puzzles of 'catsonality'; we must pick up the clues as we go along and try to use what we find to improve our relationships with our feline pets.

In any event, if you've got an aloof type, and you really want a friendly type, there may be no way out. You may just have to accept your lot, love him from afar and feel honoured if he deigns to come for a cuddle. Trying to win him round by piling on the love and encouragement may be a frustrating business that in the end merely drives him further from you. Make yourself more attractive to him to encourage his approach, rather than pursuing him; offer frequent small meals so he wants to come to you; use his weakness for warmth and comfort by making contact with him in front of the fire. (Turn off all the other heating so he will come and sit next to you!) When he does actually take the initiative, find the time to interact, but never overdo it. You may want to get another cat, hoping that number one won't leave home in disgust and that number two will be a type one cat. Careful selection will help you find a 'friendly' cat or kitten this time.

EARLY EXPERIENCE

Social relationships in kittens are most readily formed in the first two months after birth, when humans and other animals can be integrated into the group and responded to with affection, fear reactions are not yet learned and tolerance is high. Once this knowledge of another species has been acquired, the kitten often becomes resistant to change and it loses its inquisitiveness. Missing this period may mean that the kitten grows up with much less of a tendency to be sociable with cats or people. The friendliness of adult cats to people has been shown for certain to relate to the amount of handling they received as kittens.

Early handling also seems to have an effect with regard to a kitten's nervousness or boldness and, in fact, its whole attitude to life. Observers of feline behaviour found that kittens that had been handled regularly and had become accustomed to humans in their first forty-five days were more inquisitive about new objects and were more willing to approach people than those that had not been handled. Interestingly, single kittens seemed to be less nervous of strange situations and more friendly towards humans than kittens from larger litters. Perhaps in trying to compensate for a lack of litter-mates at that crucial age they made more effort to interact with other species such as us. This early time when kittens learn to respond to species other than their own is called the 'sensitive period of development' and is when we can most make our mark. Kittens handled between two and seven weeks of age are more friendly than those handled earlier or later.

Do kittens keep these characteristics throughout their lives and remain stable characters after maturity? Do timid kittens become timid cats? Do bold kittens stay confident? The answer to all these questions seems to be yes. The response of a nervous cat to stress has genetic, as well as environmental, factors. It may always over-react and never learn to compensate and to ignore everyday 'alarming' happenings despite the fact that other experiences should tell it that it won't be caused any harm. Its response does not seem to improve with handling. This tallies with the theory of the two types of cat – the excitable group may become over-anxious and agitated under challenging conditions whereas the comparatively 'laidback' type do not panic but have a more relaxed attitude to life.

So there is genetic disposition which dictates the level of friendliness, but of course it is not a simple matter. As we saw, early association with humans may determine whether the cat will be what we think of as a good 'pet' – that is, friendly and relaxed with people – so long as it has an inherently friendly character. Some experiences exert a long-term effect on how the cat develops. Early handling not only serves to socialise the kittens, it seems to speed their development and results in their being less fearful.

It may be tempting to try to 'bond' a kitten to you by rearing it yourself. However, surprisingly, hand-reared kittens or those taken from their mothers too early often develop behavioural problems and although they maybe friendly one moment, the next moment they may react aggressively to people, including their 'surrogate mother'. They may also be fearful of, and aggressive to, other cats, not learn so well or not develop physically as quickly as normal kittens raised by their mother. These behavioural deficiencies or failings may arise because we are unable to teach them 'cat speak' or to wean them behaviourally in the same way as a mother cat, and we fail to teach them adult cat communication at the same time as they are weaned off the milkbar. Their aggression with their human owners is less easy to explain, but this may be a case of muddled learning. Although we have physically weaned them off milk, we are unable to take their mother's role in weaning them behaviourally. She would have taught them to be independent and begun to distance herself soon after weaning them off milk. They have not learned how to direct their feline aggression or excitement, or how to stand on their own four feet; and they respond to their feeling of vulnerability by being aggressive.

TRAUMAS AND HORMONES

There are also two other factors which can alter a cat's behaviour, changing it dramatically from what we would call 'normal' in a cat. First are very substantial, traumatic, events in the cat's life, such as accidents, disease or injury. An encounter with a dog that causes injury may mean that the cat will not only fear all dogs but become fearful of many things it previously ignored. The encounter seems to bring on a complete loss of confidence in its ability to escape and any subsequent threat causes the cat to over-react to the situation. Confidence may return slowly, although the cat may remain fearful of dogs for the rest of its life.

Intensive nursing of ill cats may also bring about changes in character. Some previously 'untamed' cats after being intensively nursed for weeks have become very friendly and loving. Perhaps it is because they have been forced by their frailness to assume the role of kittens with their nursing humans taking the nurturing, maternal role. This reversion to kittenhood allows them to 're-learn' their relationship with humans, or one particular human at least. Too weak to follow their 'fight-or-flight' instincts, they find that the threat they have feared from humans doesn't materialise and that, even better, humans will feed them, give them warmth and companionship. Perhaps they are basically of the 'friendly' type that failed to gain enough human contact during the sensitive period of kittenhood. Now they get their first chance to enjoy close encounters with people.

Another strong influence on 'normal' behaviour is that of reproductive hormones. One may argue that these are also 'normal', but most pet cats are neutered and the cat's reproductive repertoire does not really enter the human/cat relationship. Cat breeders do, of course, keep unneutered animals, but they rarely have unneutered toms as house pets because of adult behaviour such as inherent aggression and spraying, which is triggered by male hormones. In addition, unneutered toms naturally tend to patrol a large territory, will challenge other toms and may, in many instances, be injured in ensuing fights. Abscesses from scratches or bites are common in fighting toms and the veterinary bills can mount up at an alarming rate. On top of this are the risks of frequent anaesthesia, for few cats can be treated or stitched up if not sedated or anaesthetised.

An entire tom will spray his territory more liberally and with a highly pungent urine. Such toms may also become over-excited if patted or groomed and this excitement can spill over into aggression or unwanted sexual mounting responses. He may try to grab the hand that strokes as if it were the neck scruff of a queen, and he doesn't do it gently! So cuddles and rough-and-tumble games can become a little too dangerous for comfort.

For these reasons most people have their tom cats castrated, especially if they wish to live in close proximity with them. Neutering early removes most of the desire to fight over territory and lessens the tendency to spray. Neutered cats still spray but they usually do so outside the house and with a much less smelly urine. Castration also reduces the chances of overspill of excitement during cuddling turning into aggression and it makes the cat more friendly and interactive, which is probably the biggest bonus. Other benefits are that the neutered tom is more tolerant of other cats, more amenable and playful with us and often demands more attention. He has more time to spend playing and interacting now that he is no longer driven to go out on prowl-and-patrol. Some people live with unneutered toms without encountering any such problems, but most of us don't want to risk the possibility of our relationship with them deteriorating or putting their health, in terms of fighting and roaming, in jeopardy. Castration is a very simple procedure that is completed in minutes and the cat wakes from anaesthesia to carry on as if nothing has happened.

While neutering (spaying) females does not alter their characters to such an extent, it does prevent the highly active and very noisy problem of 'calling'. During the period when the queen is receptive to males (this usually occurs twice a year but, once started, can go on every few days in cycles of three weeks until the cat becomes pregnant) she will try to let every tom in the neighbourhood know she is around. The loud calling generally takes place during the night and owners of more vocal breeds such as the Siamese need earplugs or exceptionally effective sound-proofing to sleep through such a performance. It is usually enough to convince the owners that they don't actually want a litter of kittens at all and spaying would definitely be a good idea. Like castration, removal of the uterus and ovaries is carried out very quickly under anaesthetic and the small wound requires only one or two stitches, which may be self-dissolving or need removal a week or so later.

There seems to be little difference in the relationships we have with cats of either sex. Because we neuter them young they do not develop the potential for extremes of behaviour that they would have done as entire animals, but again there seems to be less marked difference between the sexes as compared with, say, dogs. Even if neutered, dogs show very distinct male or female characteristics and interactions with their owners. So while for someone with a family and children who wants an easy-going dog to fit in I would recommend a female of the species, there is little or nothing to swing the balance when it comes to making a choice between a male or female cat, if it is neutered.

BREED CHARACTERISTICS

We've talked about genetic dispositions – the inheriting of the ability to be 'friendly' with humans and other cats and how scientists are trying to investigate more fully the mechanism of inherited character in our cats. But one of the most obvious ways in which we can see how selecting certain genes produces different types of cats is to look at the variety of physical appearance found in the different breeds. The variations are obvious – from body conformation and head shape to coat length and colour. Breeds are maintained by only allowing cats with certain characteristics to mate with other similar cats or by introducing variations in a controlled manner. Thus there is a good chance that, by using this selective method in the mating of cats, the disposition for certain behaviours may be passed on to the next generation along with the physical characteristics.

Some of the breeds originally arose naturally in geographically isolated areas while others are 'man-made', by manipulating the further breeding of the originals or by using one-off mutations to develop new types. For example, the Siamese and Angora developed in isolation from other groups of cats to form the basis of the cats we know today, but we have introduced further changes through selective breeding in order to produce colour varieties and changes in body shape. Breeds such as the Somali have arisen by cross-breeding, in this case by introducing a gene for the long hair into the Abyssinian.

The Maine Coon and Norwegian Forest Cat are the natural result of a mixture of cats which, by various means, reached Norway and the east coast of the USA. The Maine Coon was so named because early settlers thought it was a cross between a cat and a racoon!

Other breeds have been developed by breeding from one or two kittens born as mutations thrown up by nature. Examples are the Rex breeds, the Devon and Cornish Rex with their sparse curly coats, the hairless Sphynx and the short-legged Munchkin. The hairless Sphynx must surely face many problems in its daily life: obviously it will feel the cold, and while other cats spend up to a third of their waking time grooming, what does the Sphynx do? Does it still have the behavioural drive to groom and what does it do when it needs to groom itself to relax – does it just lick its skin? Its lack of whiskers could be regarded as a loss of one of its major senses because whiskers help the cat 'see' and feel its way around. Some people have even tried to breed a miniature cat but, so far, without success. Should we breed for breeding's sake just to produce more types of cat? Surely there are enough colours and personalities around already without sacrificing health and well-being for a new shape or hue.

But what we're interested in here are breed-related behaviours. Are there clear differences in the behaviour of different breeds of cat, as there are in dogs? Over the years of association with man, dogs have been bred for certain specialised tasks such as guarding, shepherding, tracking or even fighting. Others have been selected more as companions and usually have very friendly, people-orientated characters. Of course, dogs are well suited to human society because we have a very similar social set-up and they fit into our 'pack', generally at the bottom of the pecking order and usually without much difficulty. Cats, however, are solitary hunters and although they can live in groups they do not have the same strict hierarchical social structure. They have lived with man for thousands of years, but usually by their own rules and without being expected to perform particular tasks. Indeed the cat of old, which lived by its wits around the homestead and was lucky to get the occasional drop of milk or scraps of food, was selected, if it was selected for anything at all by us, for its hunting expertise in keeping the grain and food stores free from vermin. In terms of natural selection, kittens of good hunting mothers would be most sought after in such circumstances. Nowadays cats do not have to be good hunters to survive and most owners would prefer them not to hunt at all – or to be so bad at it that they fail to catch anything.

Little scientific work has yet been done on behavioural characteristics specific to breeds, but breeders and owners have noted breed tendencies, some of which do seem to be fairly clear cut. What we must remember is that the range of individual characteristics within any group of cats is very wide and that, when we generalise about a breed, any one member of that breed may, in fact, be completely different from the rest.

The potential purchaser of a long-haired cat (Persian type) knows that grooming is vital and that Longhairs are usually placid and easygoing, allowing their owners to tend to them every day. Which came first, the grooming or the docility? Presumably only docile cats were bred from as the others looked a mess because they did not tolerate such handling, or the owners ended up shredded to pieces in trying! Yet unwilling Persians are still not uncommon.

Of all the breeds, the most obvious behavioural differences can be seen in the Orientals, the Siamese and Burmese and the other breed-types in all their colour-point and coat-colour variations. Siamese are said to be extrovert, talkative, demanding, affectionate and devoted – general characteristics I expect most owners of Siamese would agree with. The Burmese is said to love human company and to be very affectionate, though persistent in demanding a lot of attention. Abyssinians are reported to be shy and fearful of strangers, and Somalis (long-haired Abyssinians) are described as good-tempered and shy. Here are some short character descriptions of certain breeds: Turkish Van – lively; Ragdoll – tolerant; Angora – fun-loving and friendly; Russian Blue – shy and quiet; Tonkinese – ultra-affectionate; Rex – playful; Korat – sweet-natured; Balinese – like the Siamese but a little quieter. Whether one can actually sum up a breed in one or two words is doubtful, but it is something to start with!

British Shorthairs, including the good old moggie, often have their character defined according to their colour. This is not so strange an attribution as it may appear since some coat colours do seem to have certain behavioural characteristics linked to them. Mention of a red head or a chestnut mare immediately makes you think of a fiery temperament – and although ginger cats are often regarded as being calm and affable, when they are roused it is easy to believe the red head factor is at work again. Only 10 per cent of ginger cats are female and male tortoiseshells are very rare. The black cat is said to be calm, but resistant to being trained.

Several surveys have recently been undertaken to try to get a little more information on the subject of breed behavioural characteristics. Bruce Fogle, author of *The Cat's Mind*, asked 100 veterinary surgeons to rank six different breeds or groupings of breeds – Siamese, Burmese, Longhairs, Somali and Abyssimans, a Shorthair and a tabby – into ten personality characteristics. *Cat World* magazine asked its readers to fill in a similar questionnaire about their cats. What emerged from both seems to back up anecdotal reports. Here's what they found.

Siamese, Burmese, Abyssinians and Foreign Shorthairs demand more attention than British Shorthairs or Longhairs. Burmese were always at the centre of activity and Siamese not far behind. On a question of confidence, again the Siamese and Burmese came to the fore. All types were reported to be affectionate but the Longhairs less interactive with their owners. Siamese, Burmese and Abyssinians also rated highly on activity, on vocal interactions, and on initiating contact. Siamese were also reported to be playful but more destructive than the others. The Persian or Longhair scored least on vocal, playful and excitability questions but rated highly with regard to 'handleability'. The domestic Shorthair or moggie came out tops in being the most friendly with other cats.

When owners were asked if their cats were happy to be left alone, their answers seemed to confirm that domestic Shorthairs are able to cope best. The Burmese had the lowest score, suggesting high dependence on human company. The Somali and Abyssinian seemed to be more aloof and less dependent on humans.

The Burmese emerges as a strong character and quotes from some owners in the *Cat World* survey sum up the breed well: 'Highly intelligent and very devoted, they are bold and love people. They also have a tendency to be very demanding, hating either being ignored or left out of the action.' Another owner adds: 'Wonderful companions, just like happy children.' I must add that, on the behavioural problems side, many cases of cats aggressive to other cats involve at least one Burmese. When such a strong character emerges in the form of a despot, it can mean terror for other cats in the house or even in the street — one or two have been known to break into other houses, eat the cat's food, spray indoors and then beat up the feline occupant! Obviously these are extreme examples and fairly rare, but it would be very unusual to hear of a Persian doing the same thing. On the other hand, if the problem concerns a breakdown in toilet training, the Persian is the most likely culprit — a finding that was revealed in one of the surveys which asked about cleanliness within the house. All breeds rated very highly but the Persian came lowest. However, as well as being beautiful, the Persian is said to be gentle, polite, relaxed, languorous, calm and sweet-natured. Some people may prefer this personality to the demanding Orientals.

From this growing bank of information on popular breeds we can now generalise somewhat with regard to characteristics and at least give potential owners a chance to choose or avoid what they wish. However, this is much more difficult in the case of some of the less popular or new breeds and with these it is best to consult the breeder and try to find out if his or her particular breeding lines possess certain characteristics.

One such breed that has worked successfully to produce a friendly pet with the wild look is the Bengal, which derives from crossbreeding with a true wild cat – the Leopard Cat. Initial worries about the docility of the breed and early generations now seem to have been resolved and it has been accepted as a recognized breed in the UK. However, the Governing Council of the Cat Fancy (GCCF), the main body which recognizes breeds within the UK, has stated that it will not recognize any new breeds which have used cats other than the domestic cat – *Felis catus*. Hence it will not recognize any breed which has been crossed with a wild species. This seems a very sensible move to try and prevent people dabbling with different types and temperaments of cat without true understanding of what the consequences might be. Surely we have enough beautiful coat colours and patterns and are still developing more within the cats we have already – let's leave the wild cats out of this and try to conserve them in their own wild habitats.

INDOOR AND OUTDOOR CATS

The aspect of our cat's personality that few of us see is its 'outdoor self'. Indoors it may be a loving kitten-like softie but, once in the wilds of the garden, Sooty can become an SAS expert in silent stalking, pouncing and killing. Claws which only ever pat you gently from slumber in the morning become unsheathed and lethal to any small victim. With all senses alert, pupils wide, ears pricked, muscles tensed ready to slip into action, it is a cat in full hunting mode and one totally alien to us – the alter ego of the 'softie' we know indoors.

We get a peep into this world when the cat plays or has a brief argument with a mate, but (fortunately) we rarely experience the true strength and power of its scratch and bite. We're unlikely to forget the incident if it does happen, though. An interesting example of this involved a beautiful tabby cat called Bomber, from southern England, which lived indoors most of the time with its lady owner. He would sit watching out of the window and when he spotted a rival in the garden he would get very excited. He would then attack the first thing that moved which, more often than not, happened to be his owner. The poor lady ended up in hospital when the deep bites on her ankles and legs became infected. The cat was thereafter allowed a much more outdoor lifestyle, so that his outdoor self didn't spoil his indoor relationship.

We forget that our pet cats are merely small versions of lions and tigers, the big cats we regard as 'savage' and to be feared. How lucky we are that the pet cat can and will modify its behaviour with us so that we can be at ease with it. Peter Neville, the cat behaviourist, has covered just this subject in his book, *Claws and Purrs*. He examines both sides of the cat's personality and how we can appreciate its more savage aspects by noting how the 'pet' can survive in almost any environment, able to adapt to live in adverse conditions using its wits and skills not only to hunt but also to scavenge and integrate with man in his society.

We have little opportunity to choose a cat without these instincts, for they are inborn in all cases. However, some cats are much keener and more successful hunters than others. Often a mother that is a good hunter will have good hunters as offspring. Some of the reasons for this may be genetic, but invariably the mother teaches her kittens well and they learn by observation. The offspring of an unaccomplished hunter may eventually teach itself, but is less likely to do so.

CHOOSING THE RIGHT CAT FOR YOU

How does all this research and knowledge help the average prospective owner to choose a moggie or pedigree cat? We know now that the first requirement is to choose a kitten that has a fundamentally friendly character; and then to handle it, and let it learn as much as possible in its early sensitive period. Thereafter we can continue cementing the relationship by using food and love and our knowledge of its natural behaviour in order to interact as much as possible.

Cat breeders can try to ensure the good temperament of their kittens that are to be sold to people who want them as pets by avoiding using nervous or aggressive cats in their breeding lines. If kittens are of pedigree strains, they are usually kept by the breeder until twelve weeks of age as this is the recommendation of the regulatory body for pedigree cat breeders and showers in the UK. However, if you have decided on a moggie you may well be able to take it home from the time it is six or seven weeks old. So long as you ensure that the kitten does not come into contact with other cats that might not be vaccinated and you have its vaccinations carried out as soon as possible before you let it go out of doors, there is little danger. Behaviourists would argue that kittens should be found homes as soon as possible after weaning – at about seven weeks – so that the new owners have the best chance to bond with the kitten in its greatest learning period. Breeders of pedigree cats can ensure that kittens are handled frequently and exposed to all types of experience – people, dogs and cats. They have to tread a finely judged line in order to ensure against infection while at the same time doing all they can to produce friendly, people-orientated non-aggressive and confident cats.

Before the time comes to select a kitten you must decide what sort of cat you desire. Do you want one that will be content to spend a lot of time on its own, that will survive fairly independently and just make an occasional appearance now and again? Or do you want one that is more interactive and will seek out your company and that of other cats?

You must decide whether you want a pedigree cat or a moggie and how much time you are prepared to spend on its physical care. Grooming Persians takes a great deal of time every day. Many people are attracted by the Oriental varieties but are alarmed by the attention these demand, or they worry about the boredom and trauma they may inflict upon such a cat if they expect it to stay alone most of the day or intend it to be kept permanently indoors. If you just want a cat that will fit into family life without much hassle, then a moggie may be your best bet. You can usually get one when it is at quite a young age and therefore much more capable of integrating into the hubbub of your crazy household and learning to take everything in its stride as it grows up.

Often it is wise to consider getting two kittens. Having two cats means they are company for each other, especially when you are out. We were faced with this particular dilemma when we had decided on the breed – Siamese – but not the number. Should it be one, which would bond to us closely, or should it be two, so that they could keep each other company? If the latter, would they bond to each other instead of us? We decided on two and have not regretted the decision. I am so often thankful that they have each other, because we just do not have the time for lots of cuddles and play and I am spared that awful feeling of guilt that a pet is not getting enough attention. The pair of them kept each other close company for the first month and then began to interact with Bullet, our moggie, when he came down off his pedestal and decided he might just like them after all. They play together but, when they get the chance or if we call them, they happily play games with us or curl up contentedly on our laps, enjoying our company then – it's the best of all worlds for all concerned. Certainly if you want a totally indoor cat it is a bit tough to expect one to stay alone all day while you are at work. Two are a much better bet.

When the time comes to choose a kitten health is, of course, a priority. Ensure the kitten has clear, bright eyes and a clean coat (check around under the tail for signs of stomach upset). It can be very difficult, in the excitement, to try to watch for signs of their personalities too. Often it's the looks of one that decides which takes your fancy, but do take a little extra time to interact with all of them if possible and talk to the breeder about the individuals in front of you. Also, ask how much they have been handled and whether they have met other animals. If you want a person-orientated, calm sort of cat, try to choose one that is playing happily and is not put off by strangers in the room, or one that comes forward to investigate you. Avoid the one cowering in the corner as it may be of a more nervous type and all your coaxing might never make it friendly. It's a pretty rough guide, but it may steel your resolve not to pick the very nervous individual that will not make a rewarding pet. Bearing all this in mind will help you to select one of the friendly, not over-reactive, type of cat mentioned earlier, one that will enjoy all aspects of a loving relationship and face life with confidence. Remember the cat you choose during those few minutes could be with you for as long as 20 years!

As soon as you get home make the kitten safe and warm. Try to steer an even course between fussing too much over it and abandoning it to its own devices. Continue to give the same food and to use the same litter as the breeder until it has settled in. Then, if you wish to change, you can do so gradually. Keep the kitten in one room to begin with (or use the cage to make introductions to other animals as outlined in chapter 7 under Introductions) and gradually let it find its feet before giving it the run of the house. Very soon, usually within a day or two, it will become confident and start to explore the home and find its way around.

There can be no guarantee, of course, that the cat will turn out to be exactly as you expected, but usually the charm of that individual will win you over and you will love him or her no matter what.

TIPS: DEVELOPING YOUR CAT'S PERSONALITY

Now we are aware of its potential, we should remember the ways in which we can obtain the best from our cat and how we can encourage the development and expression of its personality to the full.

- Decide which breed (if you want a pedigree) according to its personality traits as well as its looks.

- Go to a home where the queens and kittens live in the heart of the family so they meet people, dogs and other cats, and where they are handled frequently.

- Choose from kittens born of confident, healthy, friendly parents.

- Consider getting two kittens instead of one.

- Pick an active, sociable, calm, friendly kitten.

- Provide your new pet with lots of love, attention and new experiences.

- Use warmth and food to encourage the less interactive cat.

- Don't chase a cat and press attention on it – it will only move further away.

6

Intelligence
and Training

Are cats intelligent? Are they more intelligent than dogs because of their ability to live independently if necessary, or should they be judged to be less intelligent because they don't respond to what we ask them to do? In order to answer these questions, we must first decide what intelligence is. Is it the sum total of intellectual skill or knowledge, or is it the ability to learn new things, to associate ideas, or to distinguish one thing from another? Perhaps it is more the ability to adapt to changing circumstances and put them to best advantage. It's certainly a fairly complicated matter where humans are concerned, but what of the intellect of animals?

The ability and willingness of dogs to respond to our training instructions is often used as an example of how clever they are. Those which respond quickly or which we have directed to serve our needs, such as guide dogs for blind people or sniffer dogs hunting out drugs or criminals, are classed as intelligent. But dogs we regard as 'thick' may not be slow to learn, they may simply not be motivated to do as we ask. Hounds, such as beagles or bassets, may be extremely bright when it comes to following a trail which was left hours ago by a solitary fox running over hill and dale, but to train one to come back when called is almost impossible – yet it is one of the simplest things to teach most dogs.

What, then, of the cat? Historically our relationship has not been a direct and co-operative working one as it has been with the dog. While cats can be useful for vermin control, we have never trained them to hunt; we have merely put them in place and let the animal's natural behaviour work for us. So it would be difficult to categorise cats in the same way as we have done with dogs. Of course, many people have known 'supercats' which can open doors, bring in slippers, obey commands and generally act in what seems to be direct communication with their owners as well as 'thinking for themselves'. Perhaps intelligence would be better defined, in this context, as an ability to communicate, not just within the same species but with another species altogether – such as man – and the cat can certainly do that. And if brain size has got anything to do with brain*power*, then the cat is on a par with primates and dolphins because, relative to its body size, the cat has a brain larger than all mammals except these two.

MEASURING 'INTELLIGENCE'

How do we measure a cat's intelligence, other than by amassing and analysing anecdotal information of what certain cats have and haven't done? We have to try to devise tests which will reveal just how the cat's mind works. Using these tests, we can to some extent compare different animals but, again, many problems arise. A certain test of dexterity may suit a monkey well – it may have the required thought pathway and action already in its repertoire – but it might be less appropriate for a cat or dog that has never had to attempt such a task before. As a result the cat or dog appears slower and less understanding of what is required of it. In point of fact the animal's thought processes may have had to take large leaps to manage so well or at all and it is actually showing greater aptitude than the monkey. So a badly designed test may actually illustrate the opposite of what is happening. The whole exercise is fraught with 'ifs' and 'buts'. It's like trying to compare the skills of a carpenter with those of a computer programmer. Each has special skills which are not directly comparable or even measurable, but which in no way make one 'better' than the other.

Cats have been found to be able to learn a sequence or chain of responses to enable them to escape from a confined space. They subsequently used what they 'learned' to get out of other similar situations – we would say that this ability shows 'intelligence'.

In another experiment, a dog or cat had to remember and choose a box on which a light came on in order to get a reward. Researchers found that dogs were only capable of remembering which box lit up for up to about five minutes after the light had gone out. Cats, on the other hand, went to the correct box up to sixteen hours later – an ability to remember which ranked better than even monkeys and orang-utans in that test. If cats can form concepts, as they seemed to do with the box and the light, they may have the key elements of real intellectual ability. Their ability to link the light with the reward or to find their way out of a maze or set of obstacles is an example of what scientists call trial-and-error learning. The cat learns to perform certain types or sequences of behaviour that provide the reward of a solution to a problem, such as being shut in, or a benefit to it, such as a tidbit reward. Because the behaviour is rewarded, and the reward is given quickly enough for the cat to link it with the behaviour and not with something which has happened between the two actions, the cat learns to associate the two and is more likely to repeat that action or performance in the future.

Pavlov's salivating dogs are the result of a behavioural experiment with which we are all familiar, and an example of what is called classical conditioning, the association of one happening with the likely occurrence of another. Pavlov rang a bell just prior to blowing food powder into the dogs' mouths and, after a few such experiences, the dogs salivated merely at the sound of the bell, in anticipation of the food. Our cats do exactly the same thing when, in a flash, they appear from nowhere at the sound of the cupboard which holds their food being opened, or the clatter of their dish on the floor. We too associate certain sounds, voices, or music with past events – even those which happened twenty or thirty years ago. Amazingly enough, it is our sense of smell, though poor in comparison with that of most other animals, that most strongly evokes associations with other times or places. No doubt cats with their superior sense of smell and good memories can recognise and associate smells and happenings and thus give a wide berth to those things that resulted in an unpleasant experience.

This conditioned response is vital to the survival of animals as they learn what is dangerous or likely to harm them and what is a sign of a potentially pleasant experience. Once a response has been linked to a certain sound, smell or sight, it stays that way unless it is not reinforced or it becomes associated with something else. Thus if we move the cat's food to a different cupboard where the door makes a different click as it opens, it will soon stop rushing in every time we open the original cupboard merely to take the coffee out, and instead will begin to associate the position and sound of the new one with its feeding ritual. As animals grow up they become familiarised with regular non-threatening features of their world and do not react as strongly as they did when they were learning what is, or might be, dangerous to their survival. Thus, as it grows and learns, the cat's responses to sights and sounds are not quite so dramatic, and each excursion into the garden is no longer crammed with 'new' and exciting objects that needed instant investigation when it was a playful kitten.

If intelligence is measured by adaptability, then the cat should go to the top of the class. As a species, cats can learn to survive in almost every type of environment – desert, jungle or arctic, and with or without man's help. Looking at their lifestyle we can see that they have the ability to adapt quickly – when they want to, or have to in order to survive. They are not only adaptive but highly versatile and seem to learn quickly from their experiences. We know that as kittens they learn much by observation. I mentioned earlier how kittens became adept at pushing a lever to get food because they saw their mother successfully complete the task. Cats also learn by association very quickly. The carrying basket is associated with an invariably unpleasant trip to the vet for an injection, if only once a year, and the cat will vanish as soon as the basket is brought out of the loft! The appearance of the anti-flea spray can only mean an imminent assault by that 'spitting foul-smelling horror', another thing the cat will avoid at all costs. Flirt gives me the 'run-around-and-vanish' treatment because she observes that I am looking for her in a manner which is different from the one I use when I'm trying to find her before bedtime to put her into the kitchen. How she knows I am armed with a worming tablet I'm not quite sure, but she no doubt senses the 'intent' in my interest in her and is taking appropriate avoiding action.

RELEASING THE POTENTIAL

We know that kittens handled between two and seven weeks of age are much more person-orientated and that they generally develop into more responsive, outgoing and interactive cats if they are given lots of new experiences to react to. Like young children they take in so much, unafraid and uninhibited by thoughts of their own limitations. It is known that kittens can be weaned and begin to learn to hunt very early if milk becomes unavailable, so even at an early age they can adapt to the situation and learn a new survival strategy. A kitten kept in a non-stimulating environment during this 'sensitive period' may never learn to be inquisitive or how to tackle problems. It simply avoids unfamiliar things and circumstances. So catch 'em young for best results! The same principle is employed in training guide dogs for the blind. They are taken on as puppies by special puppy walkers who introduce them to all types of situations so they encounter as many sights, sounds and contacts with all sorts of people and animals as possible. They live with these puppy walkers in a normal lively family from the age of six weeks, immediately after weaning. In this way they learn quickly and are able to cope with most situations that arise when they become working dogs and can concentrate on their highly specialised tasks without being frightened into running away – dragging their blind owner behind them.

TRAINING TECHNIQUES

Training is shaping the activity of an animal so that it behaves in the way that is required by the trainer. It can take many forms and be carried out for many different purposes. Sometimes it means teaching 'tricks', such as the simple body postures assumed at the command 'sit', 'stay' or 'beg', or the more complex performances of a circus animal. At other times, such as in litter training, it means teaching an animal an association, whereby the animal learns to perform part of its normal repertoire in a certain place or at a certain time.

Training a cat to walk on a lead is another example. Once accomplished, it allows people who live in urban environments to take their cat out and let it explore without risk of it running under a car or being attacked by local 'wildlife'. Lead training does not only gently familiarise the cat with the sensation of wearing a harness and become used to being attached to its owner with a lead, it also means introducing the cat to the types of environment where it might later be taken for a stroll. Of course, starting when the cat is a young kitten makes the process a lot easier and the cat will come to see it as normal to go out and walk with its owner.

Some cats, especially if they are fearful, may never take to the harness or to the sensation of being on a lead, whereas confident relaxed cats are less likely to panic and get themselves into tangles and situations which only instil fear of the harness or lead. Try to make the whole exercise calm and enjoyable. Stage one involves accustoming the cat to the harness (a very soft adjustable one is best and much safer than a collar). Stop and take the harness off if the cat gets agitated and put it on again for short periods, perhaps when you feed him so that the harness has good associations. Let the cat wear it around the house and get well used to it before attaching the lead. Never drag the cat around on the lead but allow him to become familiar with the sensation of being attached to you (a lead about 2 metres long is best for training; any longer and things can get well out of control!). Reward him for taking a few steps with you – you can use food as a bribe to follow. Slowly get him used to walking with you around the house; then, when the partnership is proficient, proceed to the garden.

A cat which has become relaxed about wearing a harness in a quiet room may panic at being taken outside and having to face many new experiences, such as cars, children on skateboards or bicycles, unpredictable dogs or simply loud noises for the first time. Its training should include getting it gently and slowly accustomed to as many as possible of the things it might encounter while out walking. A cat's natural instinct is to flee from danger so you must provide security and safety if you are going to intervene in its ability to escape if it feels the need.

We can all teach a dog to sit or lie down – although, the way many are taught, this is more a credit to the dog than the teacher. But what about cats? Can they be made to 'do as they are told'? Cats are not renowned for their obedience in response to commands and for this reason are popularly supposed to be stupid, defiant or artful. However, they are, in fact, very fast learners under circumstances where their natural response tendencies are exploited. Because with dogs we can get away with breaking all the golden rules of training and they still stay with us and, to an extent, obey our commands, we probably try to employ the same tactics with cats . . . and fail dismally. But cats most assuredly are not dogs and they view life differently.

The first, basic principles of training are the same with all animals, from tigers to elephants, dolphins to horses, but successful training first involves understanding the animal's natural behaviour, for example, how it is likely to react if frightened or what it seems to 'enjoy' and what motivates it to do certain things.

The second principle is that of reward and kindness – punishment and fear actually slow the learning process. You know yourself that if you're nervous you can't think straight, let alone carry out a new task under pressure. The trainers of dolphins, for example, must always be encouraging and ensure their animals want to join in the 'game' of training. If they want the dolphins to leap in the air and over a rope, they don't just put a rope up ten feet above the water and expect the dolphins to know what to do. They begin by laying the rope on the water and rewarding the dolphins for simply swimming over it. They then raise it slightly and repeat the procedure. This is where one of the most important features of training arises: if the dolphins go under the rope, i.e. they have not understood what is required of them, the rope is immediately lowered and the process is begun again. There is never any punishment – in fact, the animal is never allowed to get it wrong. If the procedure doesn't work, it is because the trainer has gone too far, too fast. So, reward is the only outcome for the dolphin – training is a positive activity.

The same principles can be applied when a cat learns to use a cat-flap. First you leave the flap wide open and coax the cat through with vocal encouragement and tidbits so it can get used to the concept of being able to gain entry at that point. Then you gradually lower the flap using a prop to keep it up so that the cat has only to push a little and squeeze through the opening. Closing the flap a little more allows the cat to get used to pushing it and eventually he learns to push it open from the shut position. Some modern cat-flaps fit very snugly when shut and need quite a shove to break the draught-proof rubber seal. In this case, the cat will need a bit more encouragement.

If a resident cat already has the ability to use the flap, newcomers often learn much more quickly by watching him in action than from all your coaxing and bribing endeavours. Remembering what has been said about learning by associating actions with results, we realise that a cat loath to use the flap may not be being slow to learn. He may have got the knack in a very short time, poked his head out and immediately been attacked by a local rival or neighbourhood dog, a result which taught him very quickly never to put himself in that vulnerable position again and made him decide to train you to open the door for him instead! This way he ensures that you are his protection and any rival or other threat in the garden will run away and he can safely go out. So don't blame the cat for his apparent stupidity – his survival may depend on his balanced calculations of risk, not to mention his ability to use you as protector if he feels the need.

There is a method of training called clicker training which has become well known in dog-training and which also works for cats. The idea is to indicate to the dog or cat the exact action that has earned the reward. This is done with a clicker – a small plastic box which contains a flexible steel plate which makes a double-click sound when pressed. It is a very distinctive sound and the sound can be made very quickly so that the behaviour you are pinpointing can be marked very accurately. It is much better than a voice and, once the cat has made the association between the click and the reward, even the reward itself becomes less important as the animal understands that the click marks the correct behaviour and the reward will follow. As discussed above, the reward itself must suit the cat and be wanted by the cat – something expensive like chicken or prawns usually suffices!

Having established the marker signal and the reward, you can start to train. Think through what you want to do very clearly and break down the task into small stages. Choose your reward food and break it down into tiny pieces so you can use several bits in one training session. Use the clicker and reward when the cat does the right thing. Never use punishment when training. Even a mild reprimand can be off-putting – reward good behaviour and ignore the behaviour you do not want.

REWARDS

The timing of the reward is vital for success in the learning process. The giving of the reward must be done immediately and consistently after each successfully completed task. Delays of more than a couple of seconds may mean that the animal associates what went immediately before (not necessarily the task in hand) with the reward. Consistent, immediate rewards mean that the animal learns quickly and is keen to do what is required next time.

The other important thing about rewarding is that it must itself be perceived as rewarding – that is, something that the animal wants and enjoys. So, using fish for dolphins or tidbits for dogs is obviously appropriate and brings results. While verbal praise and patting will also work well with dogs, this reward is less welcome to a dolphin. But what about cats? While they may appreciate food if they're hungry, they are not really turned on by tidbits the way most dogs are. At this point we should consider which are the important motivating factors in our relationship with cats that can be seen as rewarding enough to be used in training. Warmth and attention spring to mind but, again, the cat does not always want attention. It is difficult to 'bribe' a cat! What we have to do is try to catch the cat at times when it is ready to interact and encourage it as much as possible by playing on its particular 'weaknesses' for stroking, for particular tidbits or even access to a secure warm spot – this needs a little more thought than we use when we reward the dog merely by giving him a treat.

Cats can be trained to perform many of the tricks we teach dogs, such as giving a paw or sitting, but they require time and patience – two attributes most of us are short of! It is always easier to start the training when the animal is young, when it is open to suggestion and interested in new happenings. You can teach an old dog new tricks, but it takes a bit longer. An older cat probably perceives the value of what we offer as a 'reward' (affection or a food treat) to be a lot less than that of sitting by the fire, taking a nap or going for a saunter. Deciding on and doing just what it wants is often more rewarding. Some cats may be naturals at learning and interacting, others too lazy or unresponsive to become involved.

Ann Head, who trains Arthur, the white cat which eats using its paw to scoop the food out of the Kattomeat tin, chose him because of his sweet temperament and desire to interact with her. She tries to see everything she gets him to do from his point of view. 'The gift of patience is essential; it is all done with kindness,' she explained, adding, 'you can't force a cat to work!' It's all about good relationships and enjoying the work.

PUNISHMENT?

Dogs are easy to train. Consider the basic principles of training you'll realise that we 'get away with murder' with our canine pets in our use of threat and punishment. Dogs learn despite our techniques, or lack of them. What could be less conducive to learning than a hall full of distractions such as other dogs, and a strange person standing in the middle giving orders, as happens in so many of our dog 'training classes'. The dog may do as he's told there, but not associate the learning with his everyday environment and continue with his old behaviour as before. So the 'training' of our dogs has not equipped us even to attempt to coax our cats to 'obey' – a cat would just up and leave rather than take such treatment.

Indeed, punishment is not even the opposite of reward. While rewarding increases the strength of an animal's response, punishment does not produce a decrease and the consequences of punishment may be unpredictable, especially in cats. The cat may not even associate the punishment with the misdeed – perhaps you have shaken it for bringing in a mouse or stealing the chicken off the table – but it may see your attack as coming out of the blue and entirely without reason. The more you do it, the more likely the cat is to try to avoid you in all situations, just in case you turn into that angry lunatic again.

'ACTS OF GOD'

There are times, however, when we want to prevent our cats from doing something, such as jumping on to the china shelf or the dinner table or, as my cat has just done, trying to open a newly painted window (which now has a fluffy finish!). In the case of one-off actions – such as with the window, which is normally 'on limits' – a sharp noise or hiss will stop the cat in its tracks and enable you to put it off jumping up. The 'sss' noise is most effective because cats use it themselves as a dramatic hiss, part of their repertoire to surprise or put off an opponent. Cats soon learn to abandon whatever action they have in mind if you use the 'sss' selectively and time it carefully so as to catch them just as they are intent on the act. The use of physical punishment or a lot of shouting and screaming will only make the cat fearful of you.

If you want to put a stop to a more long-term problem such as jumping on the cooker hob, a preventive measure in case it is hot on one occasion, then the 'act of God' approach can be used. The principle behind this is the association of an unpleasant (but not painful or dangerous) 'result' with the jump on to the hob. The knack is not to let the cat see it is you who is masterminding the 'happening' so he will associate it only with jumping on the hob. Thus if each time he jumps up a fine spray of water falls from the heavens, or a sharp noise makes him jump, he associates the unwelcome 'result' with the hob. A pile of empty cans carefully balanced so that they collapse when the cat jumps up on the kitchen work surface will have the same effect. With any luck the cat will then avoid the area in order to avoid the unpleasant event. The 'punishment' is actually an aversion and not associated at all with the loving owner, who is there to reassure and give back-up after such a nasty shock from that nasty hob!

The learning process involved is the same as in the wild, where the cat must learn by experience to avoid danger in order to survive. Once a cat discovers that if it passes a certain gate the resident terrier will rush out, whereupon it has to go into overdrive to get away, it will never again saunter past concentrating on other matters. The one experience is usually enough to make it avoid the area altogether or at very least to check carefully to see whether the little dog with the large teeth is out in the garden. Kittens have many such encounters every day and those that have to grow up in the wild must learn both to avoid and cope with everyday dangers to survive.

WHO TRAINS WHOM?

We are probably not aware of how much our cats have actually 'trained' us. They are not prompted by notions of dominance or punishment, but are patient enough to keep on until we learn what they want. For example, a cat that wakes its owner at six a.m. every morning by scratching on the bedroom door wants attention. Although the owner may curse and shout, he will eventually get up and let the cat in. The cat's 'reward' is a warm bed and attention – even if the owner is a little bad-tempered. The cat has 'trained' the owner and rewards him with a warm friendly purring body and a little miaow. The same applies to asking for food, rattling the window to go out, or scratching to get back in – it's the cat who trains us!

Although our cats can make their feelings extremely clear by using body language, we humans communicate mainly through the spoken word and often fail to read the body language of our own species. However, most pet owners would also like to be Dr Dolittles and talk to their animals. We can encourage or train our cat to 'talk' to us by using food and attention as rewards. By speaking in that special voice when you talk to your cat (let's face it, we all use a different tone or speak in a higher voice when we talk to our pets), you let it know you want to interact. To encourage it to reply, make friendly sounds as you prepare its meal, but only let the cat have its dish of food when it has 'spoken'. Make sure that your reaction is immediate and that the food is prepared and the bowl is in your hand ready to put in front of the cat as soon as it has spoken. The cat will learn to associate 'asking' with the reward of food or, if you stroke it only when it speaks, the reward of attention. This interaction is actually a two-way learning/training process because the cat soon becomes a double agent – taking what we have taught and using it for its own ends so that each time it wants something it only has to ask. Cats are not daft, that's for sure.

FETCH!

An interesting survey in *Cat World* magazine reported on cats who retrieved items their owners threw, not just occasionally, but many times, enjoying the game in the same way as dogs do. We train dogs, especially gundogs and those with a natural tendency to retrieve, to pick up things, bring them back and then to let us have them – so we can throw them again. Cats do bring in prey, so the carrying back is a natural part of their repertoire, but their becoming involved in a throw-and-retrieve game is quite unusual. In many of the cases reported it does not seem to have been the owner who taught the cat to play, but the cat who encouraged its human to keep throwing the item. One lady wrote: 'From an early age he decided he wanted to retrieve. He found a piece of screwed-up paper, about the size of a cigarette bent in the middle, and brought it to me. I could tell from his excitement that he wanted to play, so I threw it for him. He brought it back and that was the start of a lovely game. His piece of paper lives in the fruit bowl and when he feels like a game he will go and bring it to me. It is his party piece and he will return it to whoever throws it.'

The cats in the survey often seemed to be obsessed with the game, retrieving for hours at a time, but quite why they do it, how they learned and what they get from it is as yet unexplained. From the sample of letters it appeared that some kittens learned to retrieve by watching older 'role-model' cats in the house, although in some cases only one cat retrieved while others ignored the goings-on with the contempt only aloof cats can display. The owners thoroughly enjoyed this co-operative interactive game with their cats, although often it went on a little too long for their liking. One tired owner explained: 'From the moment I come home and settle down on the settee for the evening he is dropping toys in my lap and anxiously awaits my tossing them across the room, whereupon he leaps after them with remarkable enthusiasm, as though his life depended on the capture and retrieval of the toy. Admittedly this exercise can be quite tedious after about an hour, but I find it hard to deny him this ever so important game.'

Other owners had trained their cats in the same way as the dog (in one case it was because she was a frustrated dog-owner who treated her Siamese cat more like a dog) and in some cases the cat had grown up with dogs. One cat had been reared by a frustrated collie which was recovering from a false pregnancy and from whom the kitten had also learned to growl!

Items retrieved included balls, pieces of crumpled-up paper or foil, toys, dressing-gown cords, string and laces. The sound of his favourite toy, a foam-covered hair curler, being scratched would drive one cat called Bono into a frenzy. His owners would sit at opposite ends of the room and throw it to one another while the cat, having taken up a strategic position between them, would leap into the air to catch it. Bono's attraction to curlers ended when he discovered slow-worms, which he then proceeded to carry around the house! One cat moved on fast from pieces of paper and material to retrieving pieces of a board game while the game was in progress – a very successful, attention-gaining ploy, which is perhaps one of the reasons why cats take to such games and enjoy them so much. By learning how to initiate and then maintain contact, they can act out many of their play and hunt behaviours within the safety of their owner's presence and with his or her complete attention. These cats will initiate and finish the game and have their owners under complete control. Who said the cat is aloof, unintelligent or doesn't do as it's told? But then it's usually the cat who does the telling.

TIPS: TRAINING YOUR CAT

Be consistent, have patience and go slowly. The cat should never be allowed to 'fail'.

- Reward consistently and quickly.

- Don't feed *ad lib* if you want to use food as a reward – it will not be seen as such. If the cat has a favourite treat, use that only when training, but don't reward with food that makes him so excited that he cannot concentrate on what you are asking him to do.

- Only reward if the cat does what is being asked, or at least part of it, if you are breaking down the training into smaller steps. The reward must be associated with the action and so must be given immediately it is performed.

- Never punish. It frightens the cat and slows up the learning process.

- Aversion techniques or 'acts of God', such as a fine spray of water or a pile of empty cans that will topple over, can be used to put the cat off jumping on to surfaces where you don't want it to go.

- Always be loving and reassuring and never make threatening gestures.

- Start while your kitten is young, when it is easily aroused by everything you do and interested in joining in.

7

A to Z of Problems and Solutions

A • AGGRESSION

We all witness aggression in our cats from time to time, though, except in the cases of over-excitement during play, thankfully, it is rarely directed towards us. Cats are, of course, well provided for by Mother Nature with a good set of weapons – at the front and all four corners – but their usage to inflict injury has largely evolved to enable the cat to survive as a predator. But just what is aggression? It isn't only expressed in a predatory form when an animal hunts and kills to survive; it can occur in social situations, in territorial and other conflicts over resources and in self-defence, to name but a few. Aggression can be described as a hostile, physically damaging attack.

AGGRESSION TOWARDS PEOPLE
Aggression towards people is not common, but most of us have experienced what the Americans term the 'petting-and-biting'

syndrome. It occurs in many cats, in some after very short periods of handling, and in others after long periods of affectionate stroking when the cat suddenly attacks the hand that is caressing it. The theory is that when it is accepting handling the cat is behaving as a kitten would with its mother: it relaxes, enjoying the protection and attention. Then it seems that the adult cat, the independent, self-determining predator, takes over, and it suddenly feels vulnerable in this confined position. The cat then lashes out with a display of defensive aggression, biting and sometimes kicking, before (usually) jumping off the owner's lap, trotting a short distance away to establish a safe flight distance, and grooming itself to relax and calm down from its state of confusion.

To come to grips with this undesirable behaviour, it's best to try to predict when this state of confusion is likely to arise, and only to engage in frequent, but short, periods of contact with the cat so it never quite reaches this point. The periods of stroking can be increased gradually. During this period, it is essential not to touch the cat in sensitive areas such as on the abdomen, or around the hind legs, and in some cases it's wise to restrict the handling to stroking the back and the head.

Chapter 5 outlines research that suggests that there may be two distinct, genetically determined character types in cats – one which has a high requirement for social contact and is likely to be friendly with people, and the other which seems to have a high requirement for social (competitive) play and predatory activity and demands less friendly social contact. This second character accepts friendly social contact when it can initiate it, but may be intolerant or even defensively aggressive if pursued by owners intent on handling or petting. But with either character, actual unprovoked aggression towards owners is rare.

AGGRESSION BETWEEN CATS

While cats are not as hierarchical in their social relations as dogs, they use the same types of antagonistic behaviour to resolve conflicts over territory, such as when a cat threatens a rival with a stare and hiss in the garden, or to compete with house-mates for a favoured sleeping position or food. The social order is highly elastic in cats and there is huge variation in their sociability. Some are happy to live in huge groups, others are intolerant of every other cat. Fortunately most are capable of tolerating at least one or two others in the house and will form a loose system of social order maintained by body language, combative behaviour as required, and the acceptance of established rank.

Sometimes relations between two cats can break down for no apparent reason. If the cats were previously very friendly, then there will obviously be a better chance of repairing the damage than if they only ever tolerated each other. Unlike pack-orientated dogs, cats don't need to be part of a group – there is nothing to gain from cooperation. Hence, if cats decide not to get along, it can be hard to persuade them otherwise. Try to bring the cats together by feeding them more frequently but with smaller meals and serve the food in separate bowls brought progressively closer to each other. Food can act as a useful distraction in getting them to share each other's space, though following this suggested routine does require time and care.

In a household where cats are tolerant of each other, the chief problem that may arise comes at the introduction of a new individual or when a young cat reaches adolescence and becomes more socially competitive. Success then depends on the basic character type of the cat and its experiences as a social animal in a group of cats, as well as on relative sexual and dominance status. One example concerned a 5-year-old, by then spayed, mother Sealpoint Siamese cat and her three-and-a-half-year-old neutered son. They had never fought and had always played, slept and fed together in harmony . . . until the man of the house accidentally trod on the female's tail. She, naturally, yelled in pain but then fiercely attacked her son. He was severely frightened, of course, but although the owners separated them and gave them a few hours apart to cool off, nonetheless the female continued to attack the male on sight at every attempt to get them back together over the next few days. Here, a single traumatic incident had triggered a complete breakdown in all social relations.

In a similar case, this time involving Burmese cats normally kept permanently indoors, a total breakdown occurred between two neutered male cats after one escaped and remained outside for two days. On return the escapee may have been regarded and treated as a complete newcomer, but more likely he presented new and challenging smells in addition to his own familiar ones. Other cases, though not always of this 'diverted aggression' kind, have been triggered by certain noises, usually high-pitched sudden sounds of particular frequencies.

Treatment of fighting cats is by no means easy or quick, but good progress can be made with controlled introductions using cages to house first one cat then the other. First the aggressor is caged while the victim is allowed to walk around freely and grow used to the muted presence of the cat he has come to fear. Sharing space together again is a vital first step. Once the victim is settled, the roles are reversed. Any attempt by the aggressor to attack the caged victim obviously fails because of the protection of the cage. What you are aiming to do is to break the cycle of aggression and introduce reward for sharing space peacefully. The cats re-learn tolerance, and make calmer approaches to their caged former friend and sniff through the bars. This process can be further assisted by the owner brushing and grooming both protagonists with a solution of catnip, which helps to rebuild shared scents and makes them more attractive to each other (though catnip affects some cats and not others).

From that point of re-established tolerance between the cats, steady progress can be expected. The next step is to get the cats even closer by dividing their daily food ration into many short meals which are offered only when the cats are together, though one is fed in the cage, the other alongside. Food acts as a useful distraction to the hungry, aggressive or aroused cat. The cats can meet out of the cage only under supervision and control, perhaps on harnesses and leads, but it will always be some time before they can be allowed free access to each other and, even then, only under careful supervision by the owner for many weeks. Meetings are engineered on a very short but frequent basis as it is the cats' greeting behaviour that must be established first, so prolonged contact should not occur until later. Restoring previous good relations in such cases will probably take far longer than when a new cat is introduced to the home of a long-established resident, irrespective of character type. Like with some breakdowns in human and canine relations there may be no obvious or complete treatment path to follow – it's just a case of doing everything that seems logical and investing a lot of time and patience. If even this fails, rehoming one or other of the former friends, never to see each other again, may be better and kinder for all concerned.

B • BABIES

For many people the first thing that springs to mind on the subject of cats and babies together in one home is that 'cats sleep on babies' faces and suffocate them'. This notion is so strong that many parents-to-be get rid of a cherished cat prior to the arrival of a baby, and others never fulfil the desire to have a cat because of the perceived risk to their young children. What a pity that is when so many other parents with young babies gain so much from including a cat as part of the family.

Of course, it may happen from time to time that a cat, on discovering this new warm place, does curl up in the cot next to the baby, but the risk of harm is reduced to nil if parents follow the simple rule which applies equally for cats and dogs: *never leave any child unattended with a free-ranging pet, or where a pet may gain access to the child*, such as through an open window or cat-flap.

This rule applies to the new-born baby, the crawler and the exploratory toddler – indeed, right up to the point, usually at about six to eight years old, where a child understands when and how the family pet can be approached and handled or kindly sent away. Initially, the rule is followed in order to guard the safety of the helpless child, but later the innocent pet's safety also needs to be guaranteed.

Ideally, prospective owners of a new cat should wait until they have produced their human family. A new cat or kitten will then treat them all, adults, babies and children alike, as the norm and accept the rigours of family life more easily than an adult cat would when it suddenly has to cope with a noisy, unpredictable increase in family size. But, of course, babies (or cats, for that matter) are not always planned, so it is important to know how to treat the cat when the baby arrives.

Ensure that the baby and cat learn of each other from the earliest days. Controlled, supervised introductions should be carried out, with the cat held safely and allowed to smell the new arrival and all the things that come with it. Dirty nappies, piles of clean ones, the pram, the cot, the baby wipes, the toys – all are changes within the cat's home too, and it must be allowed to explore them and come to realise that neither the baby nor the accoutrements pose any threat to security.

Next, parents should ensure that a new baby poses no competitive threat to the cat. Most cats won't be bothered, as often they only get attention on demand and they rebuke our advances at other times.

Others, particularly interactive breeds such as Siamese or Burmese, are more demanding and may perceive a baby as competition for affection. Here it is important to be more rejecting of the cat's attempts to demand attention *before* the baby is brought home, so that your responses are not always 'on tap'. The baby's arrival is then less traumatic to the cat as it will continue to receive attention only at the owner's initiation (this practice is important for all members of the family, not just mum). If the cat then receives more attention when the baby is present, but only after the baby has been cuddled and tended to, it will perceive the baby as more of a prerequisite for attention and, importantly, that baby comes before cat. Of course, cat and baby should always be fed separately, as a cat excited at the prospect of dinner, be it its own or the baby's – food is *food*, after all – is a less controllable and potentially more hazardous creature, quite apart from the hygiene aspects. If it all sounds a bit dog-like, well, it is! A lot of competitive cats do respond well by being treated – in the nicest possible way – like dogs.

The cat may react by urinating indoors, as Bullet did on the arrival of our new baby. After anointing the duvet twice in the first week (in front of our eyes!), he seemed to get used to the baby's presence and went back to being his normal inscrutable self. So don't panic; reassure the cat and let it get used to the new arrival – most cope very well. Often cats are indifferent to a young baby after the initial curiosity value has worn off. Problems are far more likely to crop up when the baby starts to crawl. Suddenly all the cat's traditional bolt-holes under the sideboard or on the chair under the dining table, and its resting-place in the sun by the hall window, are subject to periodic gurgling invasions. Many will learn to retreat to higher and therefore safer spots, perhaps on the window-sill or atop a cupboard, but it is important to make sure that the cat has somewhere to use as a childproof refuge in every room, and especially that there is a warm, covered bed well out of reach of the rapidly developing grasp of the young child. At this stage in the child's development it is essential that parents guide the child's hand and encourage him or her to touch the cat and stroke it gently. Short frequent socialisation will start to teach the crawling child the methods of communication he or she can use confidently with the cat and will also make the cat aware of the child's right to approach, under supervision and with increasing competence at doing so.

Hygiene is always important and particularly at this stage, as crawling babies with enquiring fingers may discover the delights of the cat's litter tray. Providing one of those trays that have a top cover or placing the tray off the ground, on a table, say, may help but, better still, keep it in the kitchen and only allow the child to crawl elsewhere. Never is keeping the cat regularly wormed and the baby often washed more important!

Remember, too, another and equally important safety matter: small furry cat toys and cat furniture that may topple over (the scratching post, for example) can be very dangerous to the unco-ordinated crawler, so it would be wise to keep them out of the way while the baby is 'exercised'.

As the baby grows up and starts to toddle, his behaviour becomes more co-ordinated and predictable for the cat, which may then respond to the child much more as if he were an adult. Also many of the ground-level dangers, such as the litter tray, lose their interest as the child finds objects on the higher plane of head-height to investigate. The cat may need to find even higher escape zones, but by the time the child can reach the cat on the window-sill he should be well versed in the dos and don'ts of reacting with the cat and be more responsive to mum's instructions if the cat looks threatened.

The young child can also start to learn how to pick up and support the cat properly – here again, short frequent efforts are better, so the cat doesn't have to put up with a prolonged grasp. Also by this stage the child can begin to play a contributory role in the care of the cat, e.g. by helping mum to feed it. The socialising procedures of supervised contact should continue as often as possible so that the child not only grows up knowing how to react with the cat, but also learns to interpret the cat's body language and moods and, later, its needs. Then the child may happily become one of the next generation of cat lovers.

C • CALLING

Owners are often very worried when their female kitten starts to make loud distressed-sounding noises and to behave in a restless manner, eating less and urinating more. She may crouch in front of other cats with her tail pulled over to one side and with her rear end raised up, at the same time treading with her front paws and 'crooning'. Facially she may look angry or fearful, with ears back and pupils dilated. She will also lick the area under her tail frequently. These loud monotone vocal sounds are termed 'calling' and are a sign that the cat has reached sexual maturity and is trying to attract local toms to mate with her. It can happen from as early as three months of age or as late as up to eighteen months, and so may catch out some owners who do intend to spay their kitten but have not yet got round to it. The first season may not last long, or may go on for weeks – during which time the cat must not, of course, be allowed to go out of doors and meet entire toms.

Although veterinary surgeons prefer to spay female cats before or after a season (the period when they come on heat) usually before the first one at four or five months of age, they can be spayed while on heat and this may be the best option if the cat is continually 'calling', driving everybody mad with the noise (especially if it is one of the

Oriental breeds) and causing herself distress at not being able to go out and find a mate.

D • DEMANDING ATTENTION

Some cats can be extremely irritating in their demands and seem to change their minds or want one thing after another, one minute asking to be let out and the next demanding to be let in again. While it may be simply that the cat hasn't learned to enjoy outdoor life, usually such behaviour occurs because the cat discovers that by calling out its owner will quickly appear and offer the comfort of his or her physical presence or, better, will provide actual contact and petting. From the cat's point of view, having its owner around is better than being alone and either resolves the conflict of whether or not to go out or offers instant reassurance if it feels worried. Night-time is when a cat is likely to feel most vulnerable, but once you have been trained to get up at the sound of its calls there is no further need for it to worry! If you are, understandably, unwilling to endure the noise, it is vital that the cat's calling no longer meets with success in order for you to 'un-train' it. Endure it if you can (have a stiff nightcap before retiring and with any luck you may even sleep through it). Alternatively you could let the cat sleep in the bedroom with you, of course, but then it will probably demand that you stay awake all night or at least get up early to serve its breakfast.

E • EATING – PECULIAR HABITS

Cats are obligate carnivores – they must eat meat and are not usually interested in much else. A few do indulge in the occasional piece of cake, cheese or even chocolate, but others enjoy a rather more bizarre side-dish to their main course.

WOOL-EATERS

Why some cats should want or need to eat wool, and indeed other fabrics, is not understood. That they do is beyond doubt as many owners of clothes, carpets and furniture-covers with holes in them can testify. Although the behaviour was noted in the 1950s it was thought to be a trait restricted to certain strains of the Siamese breed. The results of a survey on the problem carried out by the cat behaviourist Peter Neville revealed that the problem is more widespread, and occurs in Siamese, Burmese and cats of mixed parentage, including moggies.

Some fabric-eaters do stick to consuming wool, and where this is the case perhaps the smell or texture of wool acts as the initial trigger to the behaviour. But the majority broaden their appetite and will consume all fabrics, from wool to cotton and synthetics. Items of clothing, preferably worn, bed linen and towels are especially popular.

When eating any fabric the cat appears to be totally engrossed in its activities and sometimes in a trance-like state. Intervention by hissing or yelling or even throwing water at the cat may cause it to stop, but often it will simply go straight back to the item or look for another in a quieter place. The cat will often take in the wool with its canine and incisor teeth but, having obtained a good mouthful, will start to grind it up using its shearing molars at the back of the mouth. The volume of fabric consumed by some cats is truly remarkable, the more so when one considers that it usually passes through the cat unaltered without causing any harm. Some do unfortunately suffer blockages in the stomach or further down the digestive system and, sadly, for a few euthanasia must be carried out because of the resulting damage. Others manage to live long healthy lives, eating wool or other fabric every day without any repercussions.

A few more are destroyed because they are too expensive in their habits for owners to live with. Some have damaged dresses worth hundreds of pounds and, in other instances, furnishings worth thousands. The average gauged from estimates in the survey is £136 per fabric-eater! Most of those owners who replied to the survey said they had learned to live with the problem and would not dream of parting with the offender; nor would they be deterred from having a cat of the same breed in the future.

One theory as to the origins of fabric-eating has it that, like wool-sucking, the behaviour is linked to a continuing infantile disposition in what are traditionally sensitive breeds which are well nurtured and cosseted by their owners.

The best treatment for cats that eat wool or fabric only in their owners' absence, but which are often clingy and dependent when they are around, may be to encourage them to grow up, to exchange any continuing infantile reactions with the owners for more adult reactions, and to keep affection to short doses at the owners' initiation. Wherever possible these cats should be encouraged to go outside in order to further their level of stimulation, reduce the importance of owners and home for activity and help establish a less dependent relationship with the owners. Often the problem does appear to resolve, though how many cats are finding fabric sources in the homes of neighbours or from their washing lines is difficult to say. Edible fabric must be made as inaccessible as possible for all fabric-eaters, and sometimes simply that denial for a few weeks causes the behaviour to cease. Direct negative conditioning by ambushing the cat with a water pistol while it is in mid-chew has helped some, though often such tactics only produce a secret fabric-eater. Indirect tactics using taste deterrents applied to specially laid towel baits can have a sufficiently dramatic effect as to put the cat off eating any fabric ever again, though the choice of deterrent is crucial. The traditional tastes of pepper, mustard and chilli or curry paste are invariably useless: they simply broaden the cat's desire for a more exotic normal diet. Aromatic compounds such as menthol and oil of eucalyptus do seem to be more successful and have reformed some ardent fabric-eaters, but need to used with care.

The best chance for curing the problem, however, seems to lie with dietary management. Most fabric-eaters have a perfectly normal, healthy intake of their proper food and their appetite is usually unaffected by having a stomach full of nylon sweater or woollen scarf. By providing a constantly available source of dry cat food in addition to offering the usual diet, the desire to eat fabric can be redirected towards more nutritional targets, and apparently without risk of weight gain. Most cats simply snack all day on the dry food and cut down voluntarily on the intake at usual mealtimes. Sometimes it helps to cease mealtimes altogether and instead simply leave a never-ending supply of dry food for the cat.

If keeping the stomach constantly active and partially full eliminates the need to eat fabric in some sufferers it may be due to the pleasurable, comforting feeling of having food in the stomach rather than to the switch to an alternative form of intake. This may explain why adding fibre, which helps pad out the volume and passage time of traditional canned diets through the cat's system, can help prevent fabric-eating. Such padding can be achieved up to a point, by adding fibre sources such as bran to the usual diet, but most cats will not accept too high a proportion. Instead, adding small lengths of finely chopped undyed wool or tissue to the diet may be more acceptable to the cat. This is, admittedly, a form of giving in, but it's a lot less expensive than letting the cat select his own fabric from the wardrobe. Other owners have resolved the problem of indiscriminate fabric-eating by providing the cat with a towel or scarf to chew at dinnertime. The cat takes a few mouthfuls of food, then chews and swallows a portion of towel – a truly bizarre spectacle, but the method is extremely effective in some cases.

With some fabric-eaters the time taken to ingest food seems to be relevant to the frequency and severity of their fabric-eating at other times. In the wild, cats would have to stalk, capture, overpower and kill their prey before consuming it. The process of consumption itself would take some time as the fur or feathers have first to be chewed or ripped in order to get to the flesh. Such gory necessities are not preliminaries to eating in the case of the pet cat, but forcing it to invest more time in processing its food may prove to be beneficial. This certainly seems to be the case with some fabric-eaters that spend much time chewing at the gristly meat and indigestible sinew attached to large bones rather than enjoying soft free meals on a plate.

PLANTS

Most cats eat more plant material than we realise, probably in an effort to obtain a quickly digestible source of vitamins, minerals and roughage. Some regurgitate grass with or without a portion of their dinner and this is believed to be a natural method of self-worming or of helping the ejection of hairballs. In general, cats are very fastidious about what they eat. Most are at liberty in gardens and houses and do not eat any poisonous plants. However, cats kept indoors permanently, and inquisitive kittens, may sample house pot plants out of a need for vegetation or just out of boredom or curiosity. Indoor cats should be provided with a tub of seedling grass sprouts to munch on, to discourage the consumption of house plants. These tubs of grass are available from pet shops and, being more attractive to cats than most house plants, are readily consumed in preference. If you do have an indoor or partially confined cat which seems fond of chewing plants, check carefully that there are no poisonous plants in the house or cat run – the Feline Advisory Bureau has a list of poisonous plants which you can send for (see Appendix).

OTHER BIZARRE DIETS

The apparently self-limiting behaviour of chewing electric cables is obviously potentially fatal to the cat and dangerous to property, and is even less well understood. Obviously the habit must be stopped at almost any cost. Encouraging play indoors with a range of toys, perhaps laced with catnip, or letting the cat spend more time outdoors may help re-divert its attentions. Make cables as inaccessible as possible and other wires as unappetising as possible using eucalyptus oil – and unplug them from the mains when not in use.

F • FEEDING

Feeding time is one of those moments when your cat will be most responsive to you. It can be used as an opportunity for learning, as well as simply providing dinner, and it is an ideal time for strengthening the bond between you. Behaviourists advise that when you take on a new cat or kitten you feed little and often, each time calling the cat and letting it encourage you to hand over the food. Later you can feed the cat less frequently or leave the food for it to eat *ad lib*.

You may notice your cat's pupils dilate with excitement and anticipation as you fill its dish. It may mew or purr and wind its tail around your legs and rub against you as it did as a kitten trying to encourage its mother to hand over her catch.

A cat that seems to have lost its appetite should be investigated. There can be many reasons, from illness to bad teeth, for this lack of enthusiasm for food. Watch its behaviour carefully; if it shows enthusiasm as you fill the bowl but stops eating after a few mouthfuls it may be that its teeth or gums are diseased and eating is just not worth the pain it causes. Veterinary surgeons can do a great deal now for tooth problems, so a visit to the surgery is called for. They may also offer advice on diet, in order both to improve general health and maintain the health of the teeth. Like people, finicky cats are often created rather than born. We make the mistake of pandering to their desire for certain foods instead of trying to feed a balanced diet, and it is often difficult to encourage a cat addicted to, say, fish to eat anything else. When changing the cat's diet to a new type or when trying to wean him off a certain flavour, try mixing in a little of the new food with the old and gradually increasing the proportion of the new content. This allows the cat to get used to the new taste and its body to become accustomed to the new food with less of a surprise to the system. Veterinary surgeons say a cat will usually eat what is offered after three days, so persist at least for this long if you are trying to switch your cat on to a more healthy diet.

Cats do lose their appetites after accident or injury and it is important to get them eating again so that the body's system is restrengthening itself rather than breaking down tissue to keep its vital functions going. Heating food to body temperature will release aromas and may encourage the cat to investigate their source. Small pieces of cooked liver, which has a strong smell, may well get the cat interested in food again. Even baby food can come in useful after a cat has had an operation as the patient can lap it up without difficulty and the flavour may be appetising. Tender loving care is vital when nursing such a cat: keep encouraging and talking to him as you offer food. Make sure new food replaces any that is rejected – cats are very sensitive to the smell of rancid food and will be put off immediately.

Cats are commonly believed to like and even to need milk – but when you consider it carefully, adult animals drinking milk made by a different species is quite extraordinary. Most animals never have the chance to benefit from this bountiful liquid after weaning from their own mother. For this reason the enzyme in the stomach that breaks down milk-sugar is not produced after weaning. While many cats have no problem in drinking milk during adulthood, in others it actually causes upset, as undigested milk passes into the large intestine and ferments, producing gas and diarrhoea. Many veterinary surgeons now advise that adult cats are not given milk. If cats are on a balanced diet there is no need for extra calcium, and water is a perfectly satisfactory fluid for them to drink.

G • GROOMING DISORDERS

Many long-haired cats are difficult to groom and quickly learn the success of using their claws to avoid such an imposition. Of course, frequent gentle grooming while they are still kittens accustoms them to the practice. Owners must try and make grooming a positive and rewarding experience rather than a battle. Choose a treat which the cat finds hard to resist (prawns are usually a good bet!) and put a tiny bit on your finger and feed the cat. As it concentrates on the food and licking your finger, groom it a little with your other hand. Put on some more prawn and groom a little more. If there are matts, either cut them out with round-ended scissors or, if they are not too tightly matted, work from the top of the hair downwards very gently. Don't pull, as you want this experience to be pleasant. Go very slowly and be patient. Keep going while the cat still has an interest in the food and stop if you get a negative reaction. Leave it for a while and then try again when all is calm. Associate the grooming with food, warmth, praise and attention. If the cat is very badly matted then you may have to take it to the vet for an anaesthetic so that the matts can be shaved or cut out.

Then, when you have it home, even without hair, start the grooming with the food. There will be no matts and therefore it will be easy to groom gently. Get the cat used to getting excited about the food when the comb (a wide-toothed metal comb is the best tool to use) comes out. Prevention is better than cure!

You can resort to a cat muzzle if all else fails. This is actually more like the hood used to calm birds of prey, or the blinkers on a horse, than the traditional dog muzzle and is made of soft material. The cat usually becomes less active, flattens down its body on to the table surface and is more tolerant of being handled and groomed.

The almost opposite problem of a cat grooming itself too enthusiastically is not very common, but is more often seen in the 'highly-strung' breeds such as Siamese, although any breed or moggie can suffer. Cats groom themselves not only to keep the coat clean and waterproof but also when disturbed or upset and may lick away the hair and even damage the skin. This self-directed behaviour is believed to produce an increased state of relaxation. With some cases, the behaviour continues at times even when the cat is apparently not anxious but prefers the relaxed state to being bored. With others, the behaviour is sporadic and occurs more in response to occasional forms of stress, such as being isolated from the owner or other house cats. When the behaviour is relatively continuous, the cat may be unable to cope with the stresses of its lifestyle, or there may be a general pressure in its environment or social group with which it cannot deal. Perhaps the reason is that it does not get on with another cat in the household or is distressed by living with a number of cats. This can be tested by isolating the sufferer for two to three weeks to see whether the behaviour continues, though of course one should not be surprised by an initial worsening of the condition due to the stress of isolation itself. Finding the source of the stress is obviously vital to treatment.

H • HOLIDAYS

Many pet owners go on holiday with mixed feelings – relief at leaving work and home pressures behind for a couple of weeks, but sadness and concern at leaving their precious cats in a cattery. Some do not take a holiday at all for this very reason, while others go where they can take the cat too. This latter option can be difficult in many ways unless the cat likes to travel and doesn't mind staying in a hotel (not many will take cats), caravan or holiday home for the holiday period. Even if the cat is accommodating, there are the obvious worries about him getting lost or wandering off and owners must ensure that he is wearing an identity tag with home *and* holiday phone numbers on it.

Some owners and cats go on holiday together every year without any problems but, for most of us, knowing the cat is in a safe, clean, professionally run cattery is the best option. If you choose your cattery carefully and check it out for yourself, then you can go away with a much clearer conscience and a lighter heart. There may be several catteries in your area, in which case it is best to visit each and ask to look around – if your request is refused, go elsewhere. The proprietors of a well-run cattery with happy residents will be only too pleased to show you their set-up. The Feline Advisory Bureau has a leaflet on what to look for in a good cattery and a list of catteries which come up to their standards of construction and management (see Appendix). Here are some things to look out for:

- An overall clean and tidy appearance.
- There should be no smell.
- Cats should look healthy and contented.
- Cats should be housed in individual units (unless they are from the same household). Cats from different households should not be able to touch each other (there should be a barrier called a sneeze

barrier, often made of perspex, between runs or a space) or go into any communal areas as this is stressful for cats and is an easy way to spread disease.

- Each cat unit should have an indoor sleeping area with a heater and an outside run.

- Each cat should have its own litter tray and bowls which it keeps for the entire stay.

- There should be a 'safety passage' or double doors on each cat unit so that cats can never just slip out and escape.

- You should be asked to prove with a vet certificate that your cat's vaccinations are up to date.

- You should be asked about your cat's eating habits, illness etc and may be asked to sign a consent form for any vetinary treatment which may be required while you are away.

- A long-haired cat should be groomed (you may have to pay more for this) while in the cattery.

Once you have found a good cattery, you will not want to go anywhere else and you will also realize how far ahead you will have to book your place. You can then enjoy your holiday. Most cats soon settle into the new routine and cope very will – in fact it's almost a personal insult to find on your return that they have been quite happy there, thank you very much!

If you are going away only for a short period, it is probably better if your neighbours or a friend visit your house a couple of times a day to check the cats and feed them. This can work well because it means that the cats do not have to be uprooted but continue in their normal environment and routines, albeit without you. If your feeders are not reliable or you are going to be away for a week or more, to be sure that the cat or cats get the attention and company they need the best solution is to have a cat sitter stay in your home. If you can't get a friend to stay, there are several companies which run a pet-sitter service and advertise in the various cat or pet magazines. Obviously, for reasons of security, you should check them out and ask for references or talk to previous customers, but it may be an option to pursue for the cat that just doesn't take to time in a cattery.

I • INTRODUCTIONS

Most cats accept the introduction of a new cat into the household without too much trouble. Plenty of contact when they are young helps cats develop the social language necessary to meet and be around others when older, and the capacity is often retained for life. Those lacking that early experience are less well equipped to deal with other cats later, and it may well be that you can only hope for a tolerant, if slightly distant, relationship rather than a loving, curl-up-by-the-fire-in-a-heap one. The other factor, one that is much more under your control, is how you introduce them. First impressions can be lasting impressions so it is essential that they meet in stages.

Smell is a vital components and mixing scents by stroking one cat and then the other can help. Do this for several days before the cats meet. It's important to introduce the new arrival to any existing residents and your normal family comings and goings in a controlled manner, so that he or she is protected from anything frightening while learning to cope. The best way to achieve this is to house the cat or kitten in an indoor cage (when unsupervised) for a couple of days, or, failing this, in its carrying basket while introductions are being made. Protected by the bars of its 'den' it will be able to look out and learn who will be sharing its life and how they behave. Containing the cat will prevent a panic runaway that might trigger a chase from other cats or dogs, and will enable you to keep everything as calm as possible. The cage also protects its occupant when other resident animals or small children make their first, and not always friendly, approaches. The cage should be housed in a well-used area such as the living room or kitchen, so that the kitten or cat is always part of proceedings and gets used to the noises and rhythms of the house.

Expect some 'swearing' when cats first meet, but be persistent as all usually sorts itself out very quickly provided excitement or a chase are not allowed to escalate out of control. Feeding resident cats outside the cage at the same time as the new inhabitant is feeding inside will help to get them progressively closer to each other. When they have reached the stage where they eat and relax in the same room, try carefully introducing them free of the cage. This usually poses no problem with a new kitten as a resident cat will not yet see it as a rival; introducing older cats may need a little more patience and, again, try distracting them with food for the first few meetings on the loose.

Bring the new arrival out frequently for exploration and play sessions on its own. Gradually and steadily allow more unsupervised freedom in one room and permit contact with other animals when they appear tolerant of the 'invader'. Then allow access to the rest of the house one room at a time, but supervise the first few exposures so that the new cat has the opportunity to learn the geography of your home without much hindrance. Make sure that in each room there are plenty of escape routes, bolt-holes under furniture and safe higher surfaces for the cat to jump on to in case it becomes alarmed. This is particularly important if you own an inquisitive dog. Replace the cage with a normal soft cat bed when the cat is well settled and needs no further controlled exposures.

J • JUVENILE BEHAVIOUR

The very reasons we enjoy our cats, their ability to relax and act like kittens with us, may be the cause of anxiety for some owners. Certain cats just don't grow up and become independent. Most pet cats switch smoothly from one mode to the other, showing kittenish behaviour with us and becoming skilled solitary hunters when outside. However, a few become so bonded to their owner that they continue to suck and knead as they did as kittens and are not able to cope without their owner's presence. You might initially enjoy the behaviour as kittens dribble and suck, often on your neck. However, it can become somewhat annoying, and indeed embarrassing, if the kitten sucks hard enough to leave red marks. This is one of the most difficult habits for owners to curtail because they must necessarily distance themselves from their cat to prevent the sucking. They usually worry that they are being 'cruel' and feel very guilty about pushing their cat/kitten away, but there really is no other way. Mother cats too have to make their offspring independent so that they will cope on their own in the big bad world.

Owners need to make themselves less available in responding to their cat's demands. That doesn't have to mean you must offer less love and attention; it simply means that these should be at the owner's instigation and control, not the cat's. It will soon learn to be more independent and enjoy the new more adult relationship, although there will no doubt be the relapses (of a less severe kind) into kittenhood that all our cats display.

K • KITTENS – WHEN DO THEY BECOME CATS?

We love our cats when they are young – you can't not enjoy their antics. But a kitten grows fast, and by the time it is six to eight weeks old it is the equivalent of an 18-month-old child. By the time the cat reaches its first birthday, it should be considered to be at the same stage as a 15-year-old youth. The comparison is valid as the cat will more than likely have had its first season and, if not neutered, will be able to have kittens.

At the age of two years a cat can be considered equivalent to a 24-year-old person and then each successive year of its life is deemed to be equal to four human years. On average the cat lives to twelve years, which makes it 64 in human terms, and a life span of twenty years (96 years) is not uncommon. This method of calculating a cat's equivalent age is more accurate than the seven-year one commonly used.

L • LITTER

The cat-litter industry in the UK is worth over £60 million per year – a remarkable fact when one considers that the majority of cats probably never use a litter tray once they are allowed out of doors, except when boarded at a cattery or when confined during veterinary treatment. Indeed, one of the reasons the cat makes such a good pet is its convenient habit of digging a hole in the garden soil for its toilet purposes and then neatly covering up its mess afterwards. But for permanently indoor cats, those kept in overnight or only taken out on a harness and leash, a litter tray has to be provided and replenished with suitable fresh litter as required in order to offer a clean toilet for this, our most fastidious of pets. While young puppies may require several weeks of careful house-training so that they learn not to soil our living areas, young kittens arrive in their new homes already completely clean in their habits. All we have to do is provide a litter tray containing a litter that can be raked and they will use it. It's something that seems to be instinctive in the kitten. Unable to urinate or defecate without stimulation of the anal and genital region by his mother until two or three weeks of age, the kitten through his own movements in exploring and tumbling out of the nest stimulates the process. Providing that material which is easy to dig in is to hand, he will soon teach himself to use that material for his toilet needs. Of course, he'll play with the litter, scatter it around and possibly actually eat it, but even at such a tender age he is also at the same time house-training himself. It's a training that will serve him for a lifetime.

The cat's ancestors evolved under semi-desert conditions and so it's no surprise that studies have shown that cats prefer a fine material to use as a latrine. While sand is ideal, and fine soil perfectly adequate, the range of modern commercially produced cat litters also suit nearly all cats. There are three main types: Fuller's earth is a clay-based and usually pale-grey granular material and probably the most popular in terms of purchase tonnage. Rising fast in the popularity stakes because of its lighter weight and odour-retention qualities is the wood-chip pellet variety, often pine scented for masking catty smells in our kitchens. There are also 'clumping', fine-grain clay litters which are claimed to offer the benefits of long active life providing that the damp clumps and solids are removed regularly to leave unsoiled litter fresh in the tray. In preference trials it has been shown that, given a choice, cats go for the litter that is granulated the finest, but the vast majority will use whichever one is provided. If house-training problems do ever arise though, offering a finer-grained litter or sand is often enough to persuade the cat to use its tray consistently again. Some cats will happily use shredded newspaper in a tray – though this may be an unwise choice of material on your part. If your cat finds your daily paper on the door-mat before you do in the morning, you may discover that he's left his little comment on your choice of newspaper.

In terms of communication, it's important to remember that the scent of urine and faeces is often used to carry messages. After urinating or defecating, a cat will rake over what he's done, often pausing now and again to sniff the area to assure itself that the right amount of scent is being given off from the deposit. In dry soil or litter he may rake frequently and endeavour to bury his deposit at a greater depth, while in a heavier or damper litter a lighter cover-up may suffice. In a multi-cat house a shy cat may try to avoid drawing attention to himself by covering up intensively, while a cat that is trying to boost his own self-confidence or is intent on leaving a dominant sign for other cats may only cover up with a thin layer of litter, or not cover up at all. Such levels of cat talk are all worthy of note, not so much in the case of the well-adjusted and house-trained cat but in individuals presenting some form of behavioural problem. Observing a cat's use of the litter tray (or especially the lack of it!) and the nature of the messages left can help enormously with diagnosis and treatment.

M • MOVING HOME

Moving home can be a very traumatic time for people and cats. Owners have many worries about how their cat will cope and how they can ensure that it does not wander off and get lost in its new environment or, if the old home is within an easy walking distance, that the cat will return there instead of accepting the new house as home. Indeed this does happen and, since often cats value their territory as much as their owners, they return to their old haunts and try to take up residence with the new people who live there. There are some things which the owner can do to help the cat over the move and to settle into its new home.

It may be wise to board particularly nervous individuals in a friendly cattery before the packing of belongings and stripping of curtains, etc. starts at the old house, and not to bring them to the new house until everything is unpacked and positioned. Outdoor cats with a wider experience of change generally cope better, but should nevertheless be kept in the new home for a week or two in order to learn the geography and smells of their new home base. When finally let out to explore their new outdoor patch and carve out a piece for themselves (temporally or spatially) from that of local resident cats, it's best if they're hungry. Starved of food for twelve hours or so, they will not wander too far from the new home and will readily respond to the call or plate-bashing that signifies 'dinner is served'. Accompanying the cat on its first few excursions into the brave new world will also help, but the cat's adaptability and survival instincts usually serve it well and it soon adopts a similar lifestyle and habits to the ones it enjoyed at the old house.

If the new home is only a few streets away from the old one, it is highly likely that in its explorations the cat will encounter old known routes. It will simply return 'home' along those routes as before and then look confused on arrival to find that all has changed. The bond with the new home is simply not yet well enough established to lure such cats. Some too are inadvertently encouraged by the new occupiers of the old base who provide food, or who are flattered by this strange cat's confident entrance through the cat-flap and willingness to set up home with them. But even when these new occupiers have been warned that the cat might return and they take deterrent action by turfing the cat out, throwing water at it and being generally unpleasant, the bond with the old centre of the territory can persist. The cat keeps returning and will only go to the new home if physically taken there by old owners collecting, or new ones delivering it. Both parties can get tired of the travelling, especially in the remarkable cases where cats have returned to old haunts many miles away.

The first step is to ensure that the new occupiers of the old house do everything to detach the cat from his old home by chasing him away and throwing water at him, and never stopping to say hello or feel sorry for him. Other neighbours, even those previously friendly with the cat, must be asked to behave similarly. The cat should be kept indoors at the new house for about a month, but if it still returns to the old abode after that it should never be taken back to the new home by a direct route. Instead make as wide a detour as possible, heading off initially in totally the opposite direction and driving, if you have that option, a good few miles before circling round and back. As a last resort, consider boarding the cat for a few weeks in a cattery as far away as possible from either home in order to scramble both its memory of the old home and its homing mechanism. But once at last at the new house, the tricks of short frequent feeds and plenty of love and attention should help build new bonds. Again, the cat should be starved for at least twelve hours before being let out, and then for the first two weeks allowed out for only one period per day, being called in within half an hour and promptly fed.

The aim is that the new home comes to be perceived as the centre of the new territory and a source of food and shelter (in contrast to the old home, where these things are denied him). It may take weeks and, in some cases, months before the cat can be allowed outside unattended. The moral of the tale is: for the cat's sake, never move to a new home less than five miles away!

If all else fails, encourage the new owners of your old house, or their neighbours, to adopt your cat permanently.

Moving home can be just as traumatic for the permanently indoor cat as obviously this involves a complete change of personal territory and can leave it feeling totally vulnerable in the new house. Slow, careful introductions, one room at a time, and lots of attention will help most cats over the stress of such upheaval within a few days. Taking pains pays dividends, but most owners would agree there are limits. Edward Lear was so devoted to his cat Foss that when he decided to move house he had his new villa built as an exact replica of the old one so that the cat was not upset by the move . . .

N • NERVOUSNESS AND PHOBIAS

The cat, like every other animal, is born with the capacity to respond to challenge and protect itself from the life-threatening dangers. The responses to such challenges are easily distinguishable from the cat's normal behaviour patterns. The reaction when startled is genetically programmed and even young kittens are seen to arch their back, erect their fur and flatten their ears when alarmed by a loud noise or sudden unfamiliar happening. During the early weeks of life a kitten, like many other mammals, will run to mum for protection immediately after being startled, if she hasn't already intervened. Gradually the kitten becomes used to commonly encountered noises or activity, especially if these are never followed by anything harmful or physically painful. In short, as the kitten grows up it learns to adapt and take most things in its stride. Most cats grow increasingly competent and confident on their way through the feline school of life and become good pets.

For others, life is one big fearful experience. They avoid any sort of challenge by hiding in a dark corner or behind the sofa. Any slight noise or movement causes the cat to run away. It never faces what it fears and therefore never learns to cope with even the mildest occurrence. Anxious cats usually have a crouching gait, low carriage of tail, and slow low movements towards a sheltered spot under a table or in a corner. From there they may peer out, still hunched and with pupils dilated in fear, trying to see but not attract attention. Having found a secure bolt-hole, the cat may avoid facing any challenge altogether and lie still, hoping the threat will disappear. Some even learn to hide themselves under the bedclothes when anxious, a reaction similar to that famous 'head-in-the-sand' method of making worries go away. Others appear withdrawn and take longer and longer to move out into open areas in the house. Whichever course the cat adopts, it is clearly distressed – and it is distressing too for the owner to observe, especially if the cat reacts in such a manner when faced with normal everyday events and never seems to get to grips with life, let alone enjoy it. The signs of its distress are clear, yet it is difficult to comfort the cat at such times without contributing further to its anxiety, because this will draw attention to it and the cat may never learn that our efforts are designed to comfort it. If we push too hard, it may resort to lashing out as a last line of defence when there is nowhere else to run. The threshold of its nervousness may fall even lower as a result. Clearly such cats are not enjoying life to the full and if the problem is left untreated it will gradually worsen.

Kittens brought up by inexperienced or incompetent mothers are more likely to be less competent themselves at dealing with change, as indeed will those relatively unexposed to challenge during the first weeks of life. When a cat displays a general nervousness it is often for this reason – having received an inadequate range of experiences as a kitten it is unable to cope with the flow of challenges as an adult. However, we can rectify this omission by providing fresh opportunities for them to learn to cope with all those happenings by exposing the cat gradually to controlled safe experiences. Nevertheless, in such cases it is rare to produce a totally competent, 'normal' cat as there are always certain things to which the cat can only truly become accustomed while still a kitten. For example, kittens brought up with puppies, or even adult dogs, rarely show nervous reactions in adulthood when dogs approach them to play. By contrast the adult cat that has had no experience of dogs is unlikely ever to be relaxed about such advances and usually retreats quickly. Some older cats can late in life learn to live with a dog, but it takes an inordinate amount of patience on your part, most only get as far as tolerating dogs and rarely will any social interaction or play develop.

You can, however, build up the confidence of many a nervous cat by providing it with a new den (a kittening pen is ideal), which you place in the main activity room of the house. By this means, normal family life goes on as usual around the cat, which is protected from physical danger by the pen. Most importantly, the cat is also prevented from running away and avoiding the challenges of changes in household members, visits of strangers, noises on the television or movement of furniture going on around it. So it has to face up to them and start to interpret what is happening. The cat is essentially 'in a womb', where it is warm, protected, and provided with all life-support systems of food, water and a litter tray, but is at the same time in a position to get used to all the things that previously induced a fearful, avoiding reaction.

Severe cases can also be helped with certain drug treatments. These can help the cat by letting it learn without panicking. Any drug treatment must of course be under the control of your veterinary surgeon.

AGORAPHOBIA

Agoraphobia is the abnormal dread of open spaces, which includes the outdoors but can also mean open areas within a room. Fortunately this is quite rare in cats.

The condition can arise from a lack of early exposure to the outdoors, or exposure delayed too long and so not coinciding with a kitten's exploratory phases of development after weaning. In the majority of cases the agoraphobic cat is unwilling to go out because of a loss of confidence caused by a single traumatic incident, such as a fight with a highly territorial local rival that may even have come into the house. Other causes include major disturbance of access points to the home base during the building of an extension or garage, a chance encounter with a stray dog in the garden or a near miss with a fast car.

The risk of encountering the rival may be present every time the cat wants to go out and its reluctance to do so thus increases steadily. It may lose the ability to cope outdoors with even mild changes such as the sound of leaves rustling or the noise of a car, even if the other cat is not on the scene. Typically the severity of the cat's reaction is exhibited with all exposures, even on trouble-free days when neither hide nor hair of another cat can be seen. It will be distressed if forced outdoors and a long, long way from enjoying its earlier outdoor lifestyle.

Treatment for agoraphobic cats is similar to that for other forms of nervousness. It involves systematic desensitisation to the stimulus through controlled exposure to the outdoors, perhaps using the carrying basket or, better still, a large secure pen in which the cat can safely spend some portion of its day outdoors. Such treatment is usually best delayed until the cause of the problem is removed. This may mean waiting until building works are completed, or even coming to an arrangement with the owner of the despotic rival about which cat is allowed out at which times, in order to avoid further conflict. Once this has been achieved, the cat can be put outdoors to re-learn that it is as safe as before the trauma and that every noise on the wind is not necessarily threatening. Owners and even willing friends should accompany it on a walk round the garden on the first few occasions, making encouraging noises to bolster its confidence. Sometimes it helps to divide the cat's meals into frequent short rations and to move its feeding area to the pen and, later, to just outside the back door. The prognosis for treatment of agoraphobic cats is often very good, depending on the amount of control over the cause of the problem that can be achieved.

Many cats and dogs are afraid of household electrical goods and the vacuum cleaner must be quite frightening from their point of view. Certainly the noise and sudden movements associated with vacuuming and even with other methods of cleaning, as well as the smells associated with polish, etc. will be challenging to a nervous character. Treatment involves desensitising the cat to these frequent normal household events in the same way as outlined above.

O • OLD CATS

Older cats, like older people, need special care. They sleep more and like peaceful warm corners in which to rest undisturbed by the hubbub of family life. Since joints stiffen up, it is important that your cat has easy access to its bed, litter tray, food and water. Its bed should be warm and preferably by a radiator, and you could perhaps provide another comfortable resting-place in a favourite sunny spot by the window. If these are above floor level, make sure that the cat can get up easily to them in stages – for example, via a chair-arm or side-table to the windowsill – rather than expecting it to leap up in one movement. Restricting the cat to one or two warm rooms may define a safer, more secure area for it, and if it is arthritic, offering several litter trays in this area will ensure it continues able to be clean. An incontinent old cat is usually a very unhappy cat, so pay special attention to its toilet needs and don't expect it to have to go too far when nature calls.

When awake, your cat will require frequent affectionate contact with you. Gentle grooming is relaxing and will help your cat to keep itself clean and maintain its self-respect. It's also important to continue to involve the old boy or girl in family life as much as you or he or she can manage. While older cats require plenty of rest, they do need to be kept involved and stimulated as well. Even though they may prefer to spend more time indoors, they should continue to be encouraged to go out, perhaps accompanied by you, on warm calm days. Your older cat may become more vocal and seek to initiate contact with you by calling out, especially at night, when it may wake up and feel alone and vulnerable. Don't be cross; it may be time to let your cat sleep by your bed so that it feels more secure at night.

Try plenty of tender loving care. It is perhaps unlikely that acquiring a younger companion will help your cat, though of course older dogs are sometimes given a new lease of life when a new pup is brought home. Cats are more sedate and less able to tolerate sharing their home and family, so instead make time to offer that extra loving care and you'll enjoy the sheer character and affection of your cat in its old age.

Consult your veterinary surgeon about the changing dietary requirements of older cats and be prepared for erratic eating habits.

They'll probably prefer frequent small light meals.
Regular check-ups will also ensure that any developing problems, such as kidney disease, which is common in old cats, can be tackled early.

P • PLAY

Kittens seem to play with anything and everything, but especially things that move. Much of their play is about sharpening reflexes and developing the hunting skills of stalk, pounce and despatch, but it is nonetheless a constant source of joy to watch. The best toys for kittens are those that allow the cat to chase. Ping-pong balls, string and tightly scrunched-up paper are cheap and as good as any special toys you could buy. Offer new environments to explore such as newspaper 'tents', paper carrier bags and boxes, which they will love to jump in. Most such games are enjoyed by adult cats too, and if your cat is to live permanently indoors you'll need to offer a steady supply to keep it interested and alert. Beware of toys which your cat could destroy by chewing as removal of swallowed portions may necessitate surgery.

Q • QUIZZICAL CATS

'Curiosity killed the cat' the saying goes, and in some cases it's true – kittens like to get their noses into everything. Bringing a new kitten home may mean that you have to reassess the safety of your home, in the same sort of way you would have to if a small child was coming to stay. Ensure all wiring is safely out of harm's way, that all potentially harmful disinfectants, bleaches and cleaning materials are well out of reach, and check that every window is shut so the new kitten can't escape while making its explorations. Remove breakable items from open shelves as high places are most popular with kittens and, yes, it probably can jump all the way up there. Remember the chimney – many kittens have ventured up these dark tunnels and become stuck, and don't forget to check that the washing-machine door has not been left open and your kitten joined the pile of washing inside.

Kittens soon grow up and out of the over-curious stage, but it's advisable to take care to start with. The more they investigate and learn the more they will want to join in with your activities. Bored, unstimulated kittens may grow up uninterested and even frightened of everyday happenings, so make good use of your kitten's quizzical nature.

R • RECOVERY FROM AN ILLNESS

The wonders of veterinary science mean that our cats can be vaccinated against many killer diseases, be cured using hi-tech drugs and equipment and given emergency care equivalent to that available to humans. However, any vet will tell you that no matter how miraculous a job he or she has done in stitching or medicating, the cat's will to recover is determined to a great extent by the quality of nursing it receives.

The cat can be coaxed into wanting to live by giving it good old-fashioned tender loving care, or it can give up the will to live merely because it becomes depressed. You as an owner of an ill cat have a great responsibility to help it pull through. Gentle talk and tending, keeping it warm and away from draughts, encouraging it to eat (warming its food or trying all sorts of tasty tidbits if it has lost its appetite), helping it to its litter tray and generally reassuring with touch can make the difference between life and death. Cats that recover from illness often form very strong emotional bonds with their owners. Perhaps they realise and appreciate the care and love that went into bringing them through.

S • SCRATCHING FURNITURE

As outlined previously, a cat scratches an inanimate object not only in order to remove the old husks from its claws to reveal sharp new points but also as a method of marking territory. A cat scratching furniture may do it for either of these reasons, or both. The sharpening aspect can be re-diverted from furniture on to a scratching post made by winding string around, or attaching bark to, a wooden pillar. The carpet-covered posts sold by pet shops may merely encourage the cat to use your carpets, so are best avoided. Place the scratching post in front of the area the cat has been scratching and after the cat uses it move the post a little distance to where it is out of harm's and temptation's way. Make sure the post is tall enough to enable the cat to stretch out to full length as it scratches. The height factor is why furniture often suits cats very well and a short post is unlikely to make a satisfactory swap for a tall upholstered chair.

If the cat is scratching to leave a scent mark, as outlined in chapter 2, it may be that it feels insecure and is trying to enhance its own confidence by having its scent around. This problem should be tackled in the same way as indoor spraying, which is another form of marking, as outlined under 'Urine spraying' in this A–Z section.

T • TOILETING PROBLEMS

Cats are famous for their cleanliness and even young kittens seem to head instinctively for the litter tray when nature calls. Make life easy during your new kitten's or cat's first days with you by placing its litter tray in the cage, or not far from its bed, but well away from its feeding area. Use the same type of litter that was provided at its previous home. Although your cat will be deterred from using the tray if you allow it to become too dirty, you don't have to hurry to attend to it. Cleaning once a day or every two days will ensure that the cat associates its own smell with the tray and perceives it as its latrine.

As the cat gets older and has completed its course of vaccinations and is allowed outside, all toilet procedures should transfer to the garden. You can assist this transition by adding some soil to the tray and moving it a little closer each day to outdoors – first nearer the access door, then on to the step outside – thereby establishing the indoors as a no-go-toilet zone from then onwards. Tip the used litter/soil mix on to a suitable patch of garden for a few days during the cat's first outdoor excursions to establish the garden as its new latrine. Remember that sick cats or those unwilling to go out in bad weather will always need a litter tray indoors. It is rare for healthy cats to make mistakes but if an accident occurs before your eyes, simply pick the cat up and place it in its litter tray. Stroke it and speak kindly; it will soon get the idea. Punishment such as 'rubbing his nose in it' is pointless, particularly after the event, and will only make it nervous and more likely to go in the wrong place again. If you don't catch the cat at it, all you should do is clean it up. If you suspect that the cat is unwell, consult your veterinary surgeon without delay and, if troubles persist, ask him or her to refer you to a professional animal behaviourist for advice.

Treatment of the toileting problem in a cat of any age will depend on the home environment and the degree of learning already established but this set of general principles may help resolve the problem.

1. Choose a suitable small room and confine the cat in a kittening pen with only enough space for his bed and a litter tray. The desire to avoid soiling the bed is an early-established one and the cat should move as far away from the bed as possible to urinate. Since he or she is confined, this behaviour will have to take place in the litter tray and within seven to fourteen days an attachment to cat litter and the tray as a latrine should be established. The cat when indoors should be kept in this cage at all times when the owner is unable to supervise. After seven to ten days of good aiming, the cat can be allowed out of the cage, but only into the room where it is kept, and the litter tray moved progressively further away from the bed. Access can then be allowed to the rest of the house one room at a time, and only under supervision, for initial introduction to each room. If the cat is ultimately to be allowed outside, then the litter tray should be moved in stages towards the back door, and then just to the outdoor step. It should be up-ended on to a suitable patch of soil when dirty so as to encourage full transfer of toileting behaviour to the outdoors, and finally withdrawn from service indoors.

2. Before allowing access to a room, all soiled areas should be thoroughly cleaned using a warm solution of a biological washing powder or liquid followed by a light scrubbing with an alcohol such as surgical spirit. The area should be left to dry before the cat is allowed supervised access.

3. A cover on the litter tray provides extra security. Use an inverted cardboard box with a hole cut in for access, or buy one of the proprietary brands of litter tray that come with an integral top.

4. The litter tray should neither be allowed to get too dirty, as this will discourage most cats, nor should it be cleaned too often. Once a day is ideal for singly housed cats (more often when many cats share a tray) as the presence of the cat's own smell on the litter will help him to recognise that the tray is his latrine.

5. Experimentation with different types of cat litter such as Fuller's earth granule types, wood-chip pellets, reusable waxed granule varieties and very fine grain litters. If the cat is to be allowed outdoors, the litter should be mixed with up to 50 per cent soil from the garden in order to help the complete transfer of toilet behaviour to the outdoors later.

6. The cat flap should be closed at appropriate times to help redefine the significance of 'indoor secure/clean zone' as compared with the 'outdoor jungle/toilet zone'. This will also help you to control the cat's access to indoors and aid supervision. It may be wise to put the cat out immediately after feeding, as cats often evacuate their bowels soon after. The cat should be encouraged to spend more time outdoors as the more it is out the more often it will have to go to the toilet in a suitable place and the sooner the garden is recognised as a suitable latrine.

7. Food should never be placed near the litter tray as this deters cats from using the tray and is often the reason behind their selecting other areas in the home for toilet purposes. By the same token, food can be placed at these inappropriate sites to act as a deterrent. Dry food is more hygienic than wet food for this purpose and will help even if the cat is usually fed a canned or meat diet. The food should be stuck down to the dish to prevent the cat from eating it.

8. Cats which toilet indoors should never be punished. Punishment after the event is pointless. Instead, and only if caught in the act, they should be picked up and placed in the litter tray, stroked and calmed. When they do use the litter tray they should be rewarded with praise and perhaps a treat.

U • URINE SPRAYING

Spraying is a normal act for most cats, be they male or female, neutered or entire. It is a territory-marking behaviour and is usually performed against vertical objects such as fence posts and bushes which rival cats may encounter. It is normally restricted to the outdoors because of competition with local pets. Cats rarely spray indoors because the home is usually secure from rivals and needs no further identification.

The indoor sprayer is therefore usually a cat that is feeling insecure or threatened and is trying to boost his or her own presence. Redecorating, moving furniture and changes in the household brought about by taking in a lodger, bringing home a new baby, or the death of a family member, for example, may all cause a cat to start spraying. The more cats that share a house, the more likely it is that at least one will spray owing to the presence of competition. Doors, curtains, windows, furniture legs and novel objects such as black plastic rubbish bags are all common targets. However, because owners provide security, spraying is rarely witnessed. The wet smelly deposits are usually discovered some time later.

Spraying and urinating are two different behaviours. Urination is conducted from a squatting position in the litter tray or outdoors in soil. Spraying is not used to empty the bladder, but to direct a small amount of urine on to a cat-nose-high position. When spraying, the cat stands with tail upright and quivering at the tip, which motion is often accompanied by a stepping movement with the back legs; and the spray is directed backwards.

When entire male cats are entering adolescence the spraying behaviour becomes particularly apparent and the smell very noticeable. The behaviour and the smell are reliably halted by castration at this stage. Entire females often spray to attract mates as they come into season, and such behaviour too is reliably halted by sterilisation. In old age, too, cats of both sexes may feel less secure even in familiar territory and begin to spray to maintain their presence against real or imagined competition.

The spraying posture can sometimes be adopted as a cat strains to urinate. This is a common response in cats suffering from FLUTD (feline lower urinary tract disease). If such signs of discomfort are observed when urinating or spraying occurs, particularly around the litter tray, the cat should be examined by a veterinary surgeon immediately. Treatment involves trying to make the cat feel secure and breaking the spraying habit. A cat should never be punished for spraying, even if caught in the act, as this will only make it more insecure and spraying more likely.

If changes have been made to the interior of the house or there are new members in the household, the cat should be denied access to most areas unless supervised. Any changed rooms should be explored with the owners present until the cat recognises the area as part of his or her indoor territory again.

Sprayed areas should be thoroughly cleaned with a warm solution of a biological washing powder or liquid followed by a light scrubbing with an alcohol, such as surgical spirit, to remove fatty deposits. The area should be allowed to dry completely and the cat only allowed back to it initially under supervision.

If the cat has access to the outdoors via a cat-flap, this should be locked shut and the cat let in and out by the owners. Cat-flaps destroy the security of the indoors and could allow other rivals in to compete with the occupant in the very place where it should feel safest. If spraying stops following this measure, a selective cat-flap could be installed later if this would be more convenient. To operate this type of cat-door the cat wears a collar equipped with an electronic or magnetic key which releases the flap lock and allows access only for the wearer.

Confining the cat in an indoor pen or one small room for short periods when it is unsupervised in the home will afford a more predictable area and help the cat feel more secure. A warm covered bed should be provided. As well as being protected by the bars of the pen, the cat will also be unwilling to spray near its bed, as keeping the sleeping area clean and dry is a principle firmly established at only a few weeks of age. If confined for more than two to three hours a litter tray should be available, sited well away from food and the bed. If the cat stops spraying you can allow access to the rest of the house one room at a time, but be sure to supervise these first ventures. The aim is for the cat steadily and increasingly to perceive the house as a safe zone shared with protective owners.

Cats rarely spray near their food, so small tubs of dry cat food can be placed at persistently used sites to act as deterrents. Dry food is more hygienic for this purpose and will help even if the cat is usually fed a canned or fresh meat diet. The dry food should be stuck to the bottom of the container to prevent the cat from eating it. Placing uncomfortable walking surfaces such as trays of pine cones or sheets of tinfoil may also deter the cat from standing and spraying at some sites.

V • VEGETARIAN CATS?

Some ardent vegetarians impose their dietary convictions on to their pets, which, if the pet in question is a dog, poses no problems. However, cats *must* have meat because their bodies cannot manufacture certain essential chemicals, such as taurine, from vegetable matter the way both humans and dogs can. Cats have evolved as hunters and meat catchers so successfully that they have never had to make these essential nutrients from lower-quality materials such as vegetation, and they eat them in the 'purer form' in meat. Dogs and humans, on the other hand, survived on plant material as well and often had to go without the meat, and therefore they evolved making best use of what they could find. So cats cannot be vegetarians, whatever the morals or beliefs of their owners.

W • WILD CATS – TAMING FERALS

There are millions of feral (domestic gone wild) cats in the world – many people first notice them when they holiday at hotels in Mediterranean resorts where the cats congregate for food. There are thought to be over one million feral cats in the United Kingdom alone and many people are involved in their welfare. Some concerned folk take food to their colonies or try to find loving homes for them. This can prove both difficult and dangerous. While a cat that has recently become a stray because it has been lost or dumped from its former home will re-assimilate into a new home fairly easily, a wild-born feral will not. As outlined in chapter 5, a cat needs to be handled by man before it is seven or eight weeks old in order for it to recognise humans as acceptable friends. If not, the cat is unlikely to relax or respond to humans, let alone be able to settle down in a normal home environment.

While it is possible to tame an eight-week-old feral kitten (and even then it may be very spitty and aggressive or very fearful) with patience and care, 'domesticating' an adult feral cat is usually a hopeless task. It may also be an extremely frightening experience for the cat as any sort of taming requires the animal to learn to live with humans in their environment. Thus the cat would need to be housed in a pen and quietly and gently introduced to everyday human goings-on. Kittens will usually respond and the younger they are the sooner they become relaxed and friendly. Older cats may simply be overwhelmed by the cage and any attempt by humans to get close. They crouch motionless or hide in a corner and if approached may become fearful and aggressive.

So if you're thinking of taking on a feral kitten and giving it a loving home, try to be sure (as far as it is possible to tell) that it's under eight weeks old and that you will have the time to be patient with it. Taking on an older feral cat may end in distress for all concerned and it may be best to neuter the adult and return it to a managed colony where it can live out its wild life under the watchful eye of regular feeders.

X • XENOPHOBIA – FEAR OF STRANGE PEOPLE

Some cats which are otherwise untroubled by changes within the house are thrown into turmoil by the arrival of visitors. The problem may not necessarily have been caused by lack of suitable experience with enough different people when young, it could also be brought about by a single unfortunate experience with a particularly noisy, frightening or unkind guest who unwittingly taught the cat to avoid all risk of repetition in the future by running away early.

The first aim of treatment is to block the cat's attempts to escape or avoid exposure to the challenge. His success in doing so, although protecting him from the danger he perceives, also precludes any possibility of him learning to cope on his own. Instead the cat is denied the opportunity to avoid visitors either by being restrained on a leash, if he is comfortable wearing a collar or harness, or by being kept in a travelling cat basket for short periods when he is to receive guests. This basket should be placed in the area where guests are invited to relax – usually the living room – before they arrive. The more willing volunteers there are the better, though the cat should first receive 'guests' that he knows, such as members of the family. They ring the doorbell instead of using the key. The cat's first reaction is the usual one of alarm and an attempt to escape, but this is prevented by the basket. Then the 'guest' enters and the cat, seeing that it is a member of the family, quickly calms down. Repetition should help the cat to begin to associate the doorbell with non-threatening arrivals.

Later, visiting guests can be asked to perform the same routine, entering the cat's room with an accepted member of the family and doing nothing more than sitting down some distance from him. It is essential that the cat grows used to their presence in gradual stages. Now his cage serves to protect him from the challenge he has avoided for so long, and he should settle quickly.

Slow progress of extremely nervous cats at this stage can often he speeded up by quelling the cat's over-reactions with a little sedative treatment such as valium under the direction of your veterinary surgeon. However, it is essential that the cat's tolerance does not become dependent on drugs. The drug is slowly withdrawn after a few days, so that his tolerance is increasingly learned and decreasingly drug-dependent. The drugs are simply a vehicle for exposing the sufferer to his problem. On or off drugs, with frequent exposure to as many different people as possible under the right conditions, the cat should perceive their arrival and occupancy of the core of his territory as being neutral. More importantly, he learns to accept their remaining inside his fight distance, the space which defines his opportunity to escape.

The next stage of treatment is a little more invasive. Now guests are asked to sit progressively closer to the cat's cage to habituate him further to their presence. This stage can only proceed as fast as the cat can tolerate and guests should certainly not attempt to touch him or even talk to him until he seems confident about their presence. Then the issue can be forced a little. Though it sounds rather unfair, the cat should be starved for up to twelve hours so that he is hungry when pressed into sharing space with his next visitor. The visitor, sitting close by his cage rather than bending down over it which would alarm him, gently proffers a small tidbit or tasty portion of a favourite food through the bars of the cage. Food cements relations far quicker than gentle voices, though the visitor and owner should encourage proceedings by talking gently to the cat while offering food. Thereafter the cat should be fed frequent short meals for the length of the visitor's stay (or patience) and as many guests as possible, as well as the family, should take part in the process. This steadily brings an increase in the cat's confidence and helps him view all guests as potential providers of food and, later, affection.

The final stage of treatment is to dispense with the cage and restrain the cat on a harness or collar and lead when guests arrive. Ask them to offer food, as before, then hold the cat firmly but gently and take him towards a single known and accepted visitor. This should be done slowly so that the cat doesn't panic as he did before he learned that visitors could be nice. The advance should be slowed down further or halted if the cat starts to look alarmed or struggles. Once alongside, guests can start gradually to stroke him. The process is complete when you yourself have stopped petting him and he is being stroked only by the guest. It may not be possible for guests to hold the cat for some time yet, if at all, because holding is an enclosing action denying escape and thus requires his total confidence. He would only be acting in the same way as a great proportion of cats if he were to restrict this honour to his family alone.

During all contact, guests' hands should initially either approach unseen from the side of the cat, or very slowly from directly in front of the cat so that they can be seen and accepted. Since the cat may regard the advancing hand as very much like a threatening paw, it should be offered very gently indeed. Bear in mind that paws have claws, which are a cat's main armoury, and therefore the approaching hand is likely to cause the cat to be apprehensive in the same way as he would be apprehensive of a swing from a cat's paw.

It may even be a good idea for the visitor to approach the cat at cat level rather than intimidating him unnecessarily by bending over him. If the prospect of crawling along the floor strikes the guest as carrying things to extremes, placing the cat – whether in the basket, on a lead or hand-held – on a table and approaching him face to face could be a dignified and less threatening exercise. Of course, it is important too that all volunteers are safe, so if the cat is at all likely to lash out defensively with a paw he should be in his basket when he receives guests for some time yet.

Y • YELLING

Odd behaviours in older cats have been coming increasingly to light in the past few years. This is probably because behaviour therapy services are now widely available and more owners of pets are encouraged by their vets to look for professional assistance and to seek help when problems crop up. We owners have changed too – we have learned to understand the behaviour of our cats more in recent times and have come to perceive greater value in the individual relationships we share with them. Increasingly, these relationships are based on respect, an attribute of all relationships that improves with the age of both the cat and the owner. Compared with, say, the all-too-ephemeral joys of owning a playful kitten, they are more deeply satisfying and enduring.

In comparison with the young cat of six to eighteen months old, or even the adult up to the age of eight, old cats present very few behaviour problems indeed. The sagacity of age that old people usually acquire through a lifetime of experience applies perhaps even more to the older cat. It has learnt how to behave in the human den, when to be part of the social scene and when not, how to let everyone know what it wants, when it wants it, and how to occupy the best spot for snoozing where it won't be disturbed or get into anyone's way. In fact, the older cat gets easier to look after as time goes on. Behaviour problems as such are far more likely to arise as signs of illness of the body 'wearing out' or just through the cat's greater need for company and reassurance. Behaviour problems in young or adult cats stem from inabilities to learn about house-training, or recover from training breakdowns caused by nervousness or conflicts with other cats. They may spray urine, defecate openly or scratch marks on furniture walls around the house. These are all signs that they are under social threat from the presence of other cats indoors or out, or, most commonly, because they perceive their den to be under some unresolvable challenge because someone has moved the furniture, or had a friend or dogs to stay. Most upset can be caused when the security of the home has been destroyed by installing a cat flap in the door, so the cat's bed, food bowl and owner's lap are suddenly available to all its rivals.

Without doubt, the most common behaviour problem in older cats is that of the nocturnal yeller. Many owners will report that their cat stared to call out to them for attention and affection since it became an older member of the feline community – and many such calls occur at night. Owners find themselves woken in the night by plaintive cries from their pet. On the first occasion, they leap out of bed to see what has upset their much-loved cat – a sudden illness or afflictions of age. When they find the cat, however, it is often just outside the bedroom door or pacing around downstairs, but looking the same picture of elderly health as it did at tea time.

Usually, owners find that the cat is not in any physical distress at all, and does not even seem to want anything in particular, such as to by fed or let out. Once they have stroked it and asked it what the matter is in a concerned voice, it settles down quietly and goes back to sleep. All it wanted was a little physical reassurance and protection in the lonely silence of the night and to be 'tucked in' again.

However, for the cat, two major events have occurred. First, the ageing animal has conceded to itself that, after years of being independent and perhaps rather aloof, even shunning attention from its owners when it wanted solitude, the time has come when it values their presence. Through feeling lonely or a little insecure, the older cat has now accepted that some of that warm, human contact could make everything right. If it can get its owners to be present, the cat can leave all vital decision-making to them for a while.

The second major event to notice is that the clever cat has now trained its owners, in true Pavlovian style and with all the skill of a champion dog handler, to respond to its demands. It has realized that, with one pitiful cry, its owners will leap to its side at any time of the day, but especially at night, to supply heaps of reassuring comfort. So, facing up to making a major decision – such as 'shall I lie next to the radiator or in my favourite sun spot?' – it will utter the same cry. Now the clever cat is assured that its owners will come and help it make up its mind by finding the most comfortable bed or offering the better option of a good cuddle and then being put where it will be most content.

Age brings its own rewards for the cat, especially once it has learned how to use its voice to full effect, when it can no longer physically attract its owners' attention by either jumping on them or rubbing around their legs. As for the nighttime problem – some owners place the cat basket next to the bed and deal with waking without having to get out from the warm covers. Others have used a baby communicator and talked to the cat over the intercom when it wakes; others leave on a radio or get a heated pad for the cat to sleep on – better than doing what one lady did every night which was to get up and replace the water in the cat's hot water bottle when it called her! Others, realizing there is nothing physically wrong with the cat, have closed their ears, put their heads under the pillows and tried to hold out so that the cat does not learn to do this every night by being rewarded by their presence – they will tell you however, just how persistent cats can be!

A cat's behaviour pattern varies according to the seasons, the weather, weekends or family activity. It will also vary according to whether food is provided in distinct meals or, as is the case with many dry diets, permanently on offer for free-choice feeding. As the pet gets older, it will tend to sleep more at those times when it would have been out hunting and generally coordinate its time in the home with the presence of the owners. Even if it doesn't interact socially with them as much as it used to, the important thing is that they are there, providing security and available for social contact if the older cat feels the need for some affection. The older the animal becomes, the more likely it is that it will look for its owners when disturbed or startled, or if it simply wakes up and finds itself along, which after all, is most likely to occur at night.

So, if your old cat suddenly becomes a nocturnal yeller, and disturbs your sleep, it may be time to let it sleep in the bedroom and derive comfort from your immediate presence if it happens to wake in the night. But if you don't want to do this, try to ignore the cries for a while. If you get up, remember that you are simply rewarding the cat's lack of confidence and ensuring that it can rely on you even when there is no real need. The longer the cries fail to pay off, the longer the cat will perhaps stay confident and independent and you will get a decent night's sleep again.

Z • ZOONOSES

Zoonoses is the scientific name given to diseases that can be passed from animals to humans. A few potential owners may be put off having a cat or enjoying a close physical relationship with it because of worries about disease. In fact, there is very little to worry about. Rabies is a potential horror, but one we do not have to consider in this country. Worms can be caught from cat faeces, but the problem can easily be dealt with by worming the cat regularly. Ringworm is not actually a worm but a fungal skin condition which can be caught from many animals, not just cats, and can be treated successfully. It is not common among pet cats but may be present in colonies of feral or farm cats.

Toxoplasmosis is a less visible problem. It is caused by a tiny organism which can be excreted in the cat's faeces. Affected cats show little or no sign of infection and most people who become infected suffer flu-like symptoms at worst. In fact, the organism is more likely to be contracted by eating uncooked or partially cooked meat than from contact with cats. Pregnant women should be especially careful to wash their hands after touching cats, and get someone else to clean the litter tray, as the disease is a risk to the unborn baby. There is really negligible risk of catching diseases from our cats if we keep them wormed and undertake the usual hygiene precautions. The health benefits of pet-keeping are well known, and the small risk of catching a disease is far outweighed by the years of fun and companionship a cat will undoubtedly bring.

GCCF: Governing Council of the Fancy
4-6 Penel Orlieu
Bridgewater
Somerset
TA6 3PG

Cats Protection
17 Kings Road
Horsham
W Sussex RH13 5PP
Tel 01403 61947

Association of Pet Behaviour Counsellors
PO Box 46
Worcester WR8 9YS
Tel 01386 751151

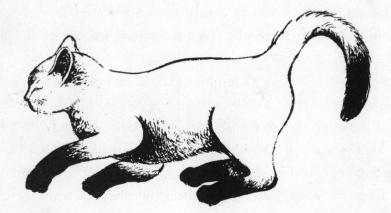